The Surface of the Earth

FOUNDATIONS OF EARTH SCIENCE SERIES

A. Lee McAlester, Editor

Structure of the Earth, *S. P. Clark, Jr., Yale University*

Earth Materials, *W. G. Ernst, University of California at Los Angeles*

The Surface of the Earth, *A. L. Bloom, Cornell University*

Earth Resources, *B. J. Skinner, Yale University*

Geologic Time, *D. L. Eicher, University of Colorado*

Ancient Environments, *L. F. Laporte, University of California at Santa Cruz*

The History of the Earth's Crust, *A. L. McAlester, Yale University*

The History of Life, *A. L. McAlester, Yale University*

Oceans, *K. K. Turekian, Yale University*

Man and the Ocean, *B. J. Skinner and K. K. Turekian, Yale University*

Atmospheres, *R. M. Goody, Harvard University*
and J. C. G. Walker, Yale University

Weather, *L. J. Battan, University of Arizona*

The Solar System, *J. A. Wood, Smithsonian Astrophysical Observatory*

The Surface of the Earth

Arthur L. Bloom

Cornell University

Prentice-Hall, Inc., Englewood Cliffs, New Jersey

 Printed in the United States of America. Library of Congress Catalog Card No. 68–10422.

Design by Walter Behnke

Illustrations by Vincent Kotschar

PRENTICE-HALL INTERNATIONAL, INC., *London*
PRENTICE-HALL OF AUSTRALIA, PTY., LTD., *Sydney*
PRENTICE-HALL OF CANADA, LTD., *Toronto*
PRENTICE-HALL OF INDIA PVT. LTD., *New Delhi*
PRENTICE-HALL OF JAPAN, INC., *Tokyo*

Current printing (last digit):
10

FOUNDATIONS OF EARTH SCIENCE SERIES

A. Lee McAlester, Editor

C

Foundations of Earth Science Series

Elementary Earth Science textbooks have too long reflected mere traditions in teaching rather than the triumphs and uncertainties of present-day science. In geology, the time-honored textbook emphasis on geomorphic processes and descriptive stratigraphy, a pattern begun by James Dwight Dana over a century ago, is increasingly anachronistic in an age of shifting research frontiers and disappearing boundaries between long-established disciplines. At the same time, the extraordinary expansions in exploration of the oceans, atmosphere, and interplanetary space within the past decade have made obsolete the unnatural separation of the "solid Earth" science of geology from the "fluid Earth" sciences of oceanography, meteorology, and planetary astronomy, and have emphasized the need for authoritative introductory textbooks in these vigorous subjects.

Stemming from the conviction that beginning students deserve to share in the excitement of modern research, the *Foundations of Earth Science Series* has been planned to provide brief, readable, up-to-date introductions to all aspects of modern Earth science. Each volume has been written by an authority on the subject covered, thus insuring a first-hand treatment seldom found in introductory textbooks. Four of the volumes—*Structure of the Earth, Earth Materials, The Surface of the Earth,* and *Earth Resources*—cover topics traditionally taught in physical geology courses. Four more volumes—*Geologic Time, Ancient Environments, The History of the Earth's Crust,* and *The History of Life*—treat historical topics. The remaining volumes—*Oceans, Man and the Ocean, Atmospheres, Weather,* and *The Solar System*—deal with the "fluid Earth" sciences of oceanography and atmospheric and planetary sciences. Each volume, however, is complete in itself and can be combined with other volumes in any sequence, thus allowing the teacher great flexibility in course arrangement. In addition, these compact and inexpensive volumes can be used individually to supplement and enrich other introductory textbooks.

Contents

Introduction

The face of the Earth is an honored subject of scientific inquiry. Since the earliest days of the shepherd philosophers, men have meditated on the dubious durability of the "everlasting" hills. In what may be the oldest textbook on landscape evolution, it is said: "Every valley shall be exalted and every mountain and hill shall be made low; and the crooked shall be made straight, and the rough places plain."

How has the surface of the Earth achieved its present form? Does it change perceptibly within the lifetime of a man, or within the time of recorded history? What forces shape landscapes? These questions have been asked by thoughtful men of every century, and from the time when modern sciences evolved out of the natural philosophy of the Renaissance, observation and measurement of the Earth's surface have been highly regarded scientific activities.

The history of landscape study is an integral part of the history of geology and geography. In the late eighteenth century, Dr. James Hutton of Edinburgh expounded a protracted *Theory of the Earth, with Proofs and Illustrations* in which he argued for the constancy of natural forces through long geologic time, in opposition to his contemporaries who generally held that changes in the Earth are due to catastrophic events. Hutton's concepts, which evolved into what is called the "uniformitarian doctrine," were based on observations in approximately equal parts of rocks and landscapes. Hutton's young friend and champion, John Playfair, wrote a lively defense of the Huttonian theory in 1802, entitled *Illustrations of the Huttonian Theory of the Earth,* which stated many of the principles we still accept and elaborate upon. The origin of valleys as the work of the streams that flow in them; the adequacy of ocean waves to shape coasts; the constant renewal of the soil by the weathering of rocks; the transportation of huge boulders by Alpine glaciers: These observations by Playfair could and do form the outline of any modern textbook of landscape interpretation.

The science of landscape study, or *geomorphology,* is part of two related disciplines. In most European countries, the subject is studied by physical geographers as part of the science of geography. In the United States, geomorphology is usually regarded as a branch of geology. In any event, the name we give to an academic subject is far less important than its content.

Geomorphology is not just a scientific discipline. Every national park in the United States, and indeed every great scenic attraction of the world, has as its foundation a uniquely spectacular landscape. In these days of extensive travel, every tourist should be a geomorphologist. The only difference is a question of degree: A tourist looks at a landscape, but a geomorphologist sees it. The far side, the deep interior, or the past history of a mountain is as visible to a trained mind as the panorama is to the eyes.

As an academic discipline, geomorphology experienced spectacular growth from about 1890 to about 1930. Two decades of relative quiet followed, for in the 1930's, a new awareness of the human world displaced the study of the physical environment by geographers. Especially in the United States, man with his crops, cattle, commerce, and cities became the chief subject of geography. As a geologic subject also, geomorphology failed to keep pace as other fields of geology advanced. Much of the fault here must be laid to the persisting trend in geomorphology toward qualitative, subjective description, while other fields turned toward the quantitative and experimental approach.

W. M. Davis (1850–1934), the acknowledged master geomorphologist, developed the technique of "explanatory description" of landscapes to a high state of perfection. His principle was that if one could describe the evolutionary history of a landscape in terms of three factors—structure, process, and time—then the description was complete. Few experimental studies on processes were conducted, for the descriptive system was deceptively adequate. Besides, since experimentation in the processes of landscape evolution is so difficult, it was largely ignored. Intuitive reasoning outraced experimental verification. Davis even complained that "there are not enough kinds of observed facts on the small earth in a momentary present to match the long list of deduced elements of the scheme."

After Davis's death, geomorphologists largely devoted their efforts to either describing more landscapes within the framework of the established scheme or devising alternate descriptive schemes. Some progress toward quantification and experimental verification was made, however, particularly in the study of soils. Then World War II created demands for new kinds of geomorphic analysis as well as for new equipment and techniques to do the analyses. For example, interpretation of aerial photographs was developed to a high degree of accuracy and efficiency by all nations involved in the war.

Accurate quantitative interpretation of beaches and coasts was especially required by the development of the amphibious assault as the major military tactic of World War II, resulting in an enormous research effort to understand the processes that shape coasts. Surf forecasts, seasonal or daily changes in the nearshore ocean-bottom profile, and the extent to which beach sand could support motor vehicles became vital pieces of intelligence for the military forces. Quantitative studies of nearshore marine processes produced a voluminous literature of technical memoranda and research reports that is still being consolidated and summarized. In postwar years, beach studies led the trend to quantification and experimental verification. The questions "How fast?" "How much?" and "By what means?" that military commanders had asked of beach specialists and aerial-photo interpreters now were asked about all landscapes. Soil erosion and river floods received new attention. A notable advance was that the sediments of a river or a beach began to be considered as a vital part of the geomorphic setting.

The postwar trend toward quantification grew so rapidly that in the 1950's geomorphology threatened to become a branch of topological mathematics. Landforms, sediment grain-sizes, the hydraulic properties of streams, and flood frequencies and durations were all reduced to statistical probabilities. A new geomorphic jargon developed, borrowed from other sciences. Authors described rivers in terms of entropy increase, steady-state systems, random walks, least work, and Monte Carlo theory. Geomorphology became a subject studied by teams of geologists, hydraulics engineers, meteorologists, and sedimentologists. Glacier ice was regarded by some as a material whose thermal and physical properties were to be analyzed in physics and engineering laboratories, rather than as a landscape to be climbed over with ice axe and crampons. The rush to disavow "classical" or "Davisian" geomorphology was so intense that at least one leading geomorphologist studied an area with the preconceived intention of disproving previous Davis-style explanatory descriptions. The dangers of prejudgment are as real in science as in any other area of human activity. It may be both a curse and a virtue that each scientific generation must not only destroy ancient idols but also erect new ones, so that progress may be made.

We are now at the threshold of a new golden era of geomorphology. The new concepts are settling in as useful tools, but at the same time we are witnessing a healthy return to old principles. For example, in Chapter 4, you will find summarized a theoretical explanation for the downstream decrease in slope of rivers based on thermodynamic principles, first published in 1964. The new theory predicts the same sequence of valley and channel shapes that were intuitively deduced by W. M. Davis and other nineteenth-century geomorphologists. The difference is that now we understand *why* landscapes change instead of being content with an assertion that they do change. Experiment is catching up with intuition.

As one of a series of short books, this volume must intermesh with the others of the series, yet be sufficiently complete to stand alone. The scheme of this book, and the areas of mutual concern with the other books of the series, can be summarized as follows. We will assume a terrestrial landscape made of rock (Ernst, *Earth Materials*), built by internal forces of the Earth (Clark, *Structure of the Earth*) acting through time (Eicher, *Geologic Time*) in the presence of life (McAlester, *The History of Life*). The landscape is the result of the reaction of rocks to atmosphere (*The Atmosphere*) and to the force of flowing water, under a bath of solar energy (Wood, *The Solar System*). The landscape is degraded, and the waste products are carried downhill to the sea (Turekian, *Oceans*) where they accumulate as sediments (Laporte, *Ancient Environments*) until internal forces convert them to new rocks and raise them as new lands.

We will review in successive chapters the energy supply that is available and the processes that act on rocks at the surface of the Earth; the movement of debris downhill, primarily by flowing water; the sequential evolution of a landscape as the degradation proceeds; the accumulation of sediments at the edges of the land; and the remarkable special effects of glaciers in changing a landscape.

Two over-all themes will be repeatedly stressed: (1) Landforms tend to develop in equilibrium with the processes that shape them (and a corollary: the processes can be inferred from correct interpretation of the forms), and (2) sediments in transport through eroding landscapes serve as shock absorbers to peak energy inputs and permit balanced states to be developed and maintained in systems of declining energy supply. The meaning of these themes should become clear in subsequent pages.

1

Energetics of the Earth's surface

Every particle of rock at the surface of the Earth has many forces acting on it. These forces come from a variety of energy sources, including the Sun, the Moon, and the hot interior of the Earth. In order to understand why and how landscapes change, we must first consider these forces of change and the sources of energy that drive the forces.

Gravity: The Leveler

Every mass has an inherent attraction for every other mass. This weak force called gravitation is an intrinsic property of mass. Although its nature is yet to be defined, we know something about its operation. We can assume, for one thing, that the gravitational force between two large objects operates as though the entire mass of each object were concentrated at the center of the mass, or the *center of gravity*. Objects near the surface of the Earth are strongly attracted toward the center of mass of the Earth, and because all objects at the surface have small masses as compared to that of the planet, we usually regard the force of gravitation as a tendency for an object at the surface to fall or accelerate toward the center of the Earth. The gravitational force is reduced somewhat, however, by the tendency of objects on this rotating spheroid to maintain straight-line motion and hence be lifted from the surface (the centrifugal effect), especially near the equator. Gravity, as we commonly use the word, refers to the net force of gravitation as reduced by the centrifugal and other lesser effects. The result of all the variables is that an object that weighs 189 pounds at the poles weighs only 188 pounds at the equator.

If the Earth were entirely covered with a free-flowing liquid, the surface of the liquid would be an oblate spheroid, slightly flattened at the poles and bulged at the equator. This ideal surface, called the *geoid,* would be in perfect equilibrium with all gravitational and rotational forces. Sea level, over the oceanic 71 per cent of the Earth's surface, closely approximates the geoid, and because the geoid is a reference surface of gravity equipotential, we should also visualize it under all the lands, as an extension or projection of sea level.

All water that falls as rain or snow on the 29 per cent of the Earth's surface that projects above the sea tends to move downhill by gravity, back to the ocean. Thus, every raindrop that strikes the ground has potential energy proportional to the product of its mass and the altitude above sea level at which it strikes. The few localities where dry land extends well below sea level (Death Valley, California, at −282 feet; the Dead Sea at −1286 feet; and others) are exceptions to the rule that sea level is the limit for downhill flow of water. However, each of the basins that extends below sea level is in an arid region where few raindrops ever fall. In fact, such basins never last very long in humid regions, for they are soon filled with water to overflowing and sea level again becomes the ultimate downward limit of water flow from the basins.

Rock particles as well as raindrops are attracted toward the center of the Earth by the force of gravity. When rock material moves downslope under the influence of gravity, but without a transporting agent such as flowing water or glacier ice, the process is called *mass-wasting.* This is the subject of Chapter 3. Mass-wasting includes not only such spectacular events as landslides and avalanches, but also the slow, barely perceptible process of soil creep.

Gravitational Forces of the Moon and Sun: Tides

The Moon and Sun are the only two celestial bodies close enough to the Earth to affect surface processes by their gravitational forces. The mass of the Moon is only 1/81 that of the Earth, but it is sufficiently close to the Earth that the two masses revolve monthly around a common center of mass within the

Earth, about 3000 miles from the center of the Earth or about 1000 miles below sea level, on the line between the centers of the Earth and Moon. The orbit of the Earth around the Sun is actually the path of this common center of mass of the Earth-Moon system (see Fig. 6–3). In addition to its daily rotation around a polar axis, the Earth also revolves monthly around the center of the combined masses, much as you shift your weight back onto your heels when you swing a heavy weight in a horizontal circle.

The principal result of the lunar gravitational attraction, of course, is to create tides. The Sun, of enormously greater mass than the Earth (330,000 times as large) is so much farther away than the Moon that its tidal effect on the Earth is only 46 per cent as great as that of the lunar tide. Together, these celestial neighbors raise tides in both the surface water of the Earth and in rocks as well. It is estimated that a tidal bulge of 30 centimeters (about 1 foot) is raised in the solid rock of the Earth on the side facing the Moon.

The tides in the water that cover so much of the Earth are better known and of greater geomorphic significance than the tides in the solid rock. The range of the oceanic tides varies greatly with latitude, season, phase of the Moon, and many other factors, but the cyclic changes of water level caused by tides permit the waves of the oceans to attack coasts through a vertical range of many feet. Thereby the energy input of the waves is expended over a greater surface area of coastal rocks than it would be if sea level were fixed, and the gravitational forces of the Moon and Sun become significant factors in the shaping of terrestrial landscapes. The role of tides in shaping coasts is described at greater length in Chapter 6.

Internal Heat

The average heat flow outward through each square centimeter of the Earth's surface is about 1.25×10^{-6} calories per second, or about 40 calories per year. This small amount of heat could annually melt a layer of ice only 0.5 centimeter (0.2 inch) thick. If you consider that a layer of ice that thick can be melted easily from the ground by bright sunshine on a winter day with the air temperature barely above freezing, you can better appreciate the very small amount of energy supplied to the Earth's surface from its interior. In fact, the average solar radiation received yearly at the surface of the Earth has about 4,000 times as much energy as the average annual heat flow from the interior, and direct overhead sunlight in clear desert or mountain air may briefly supply 17,000 times as much energy. Therefore, with the two exceptions that follow, internal heat of the Earth can be generally disregarded as a source of energy in the alteration of landscapes.

For one exception, internal heat is an important factor in deforming the

Earth's rocky crust. Most large-scale landforms such as mountain belts and island arcs probably are related to concentrations of heat within the Earth.* Certainly volcanoes, which build some of the world's most impressive landscapes, are related to abnormally high heat flow from the interior of the Earth. Whether the excessive heat flow is a cause or an effect of the volcanic activity, however, is a problem not yet solved.

There is also a second aspect of internal heat flow that may have a bearing on geomorphic processes. Temperate glaciers (in which ice and water coexist—see Chapter 7) are perfect thermal insulators, for they are at nearly the same temperature throughout their thickness, and heat cannot be conducted without a temperature gradient. If the heat flow beneath a temperate glacier is comparable to the world average, a layer of ice 0.2 inch thick will be annually melted from the bottom of the glacier. This amount of melting would be insignificant to a glacier a mile or more in thickness, except that the presence of a thin film of water at the base of a glacier radically changes the speed of glacier motion and the action of a glacier on its bed. A cold glacier, frozen firmly to the rock surface beneath, moves by plastic deformation of the ice crystals that form the glacier. It may not "slip" over its bed at all, hence erosion is minimal. On the other hand, a glacier separated from the rock floor by a thin film of water will slip readily on the ice-rock interface, and may be capable of much erosional and depositional work. This aspect of internal heat flow as an influence on geomorphic process, which has not been sufficiently noted in the past, is analyzed further in Chapter 7.

Solar Radiation

A major goal of geomorphology is to understand how the immense energy of solar radiation is converted into mechanical work that shapes landscapes. We can visualize the process in terms of a "geomorphology machine" (Fig. 1–1) in which a steam engine fired by the Sun powers a series of fans, saws, files, grinders, and hydraulic jets to reduce the landscape. In fact, the analogy with a steam engine is a good one, for it is the abundant water at the surface of the Earth that converts solar energy into mechanical work.

It has been determined that the solar energy intercepted by a surface one square centimeter in area at a distance from the Sun equal to the Earth's mean orbital radius is about two calories per minute. This value of two calories per square centimeter per minute is called the *solar constant.* The whole Earth intercepts a quantity of solar energy equal to the product of the solar constant and the area of a circular disk equal to the cross section of the Earth. The cross section of the Earth equals πr^2, where r = 6,371 kilometers, and from this

*The origin and dissipation of internal heat are discussed by S. P. Clark in another book of this series, *Earth Structure.*

FIGURE 1–1 *The "geomorphology machine" (after the style of Rube Goldberg).*

area the daily or yearly receipt of solar energy by the Earth can be calculated easily. However, the Earth is not a disk but approximately a rotating sphere, with a surface area of $4\pi r^2$ that is four times as large as the equivalent disk cross-section. Thus, the average input of solar radiation on each square centimeter of the upper atmosphere is only one-quarter of the solar constant, or about 0.5 calorie per minute.

Of the radiant energy reaching the Earth, an estimated 35 per cent is reflected directly back into space. About two-thirds of the reflection is from our cloud cover. Atmospheric dust and haze reflect most of the rest, and a small amount is reflected from water, rock, and ice surfaces. The remaining 65 per cent of the incoming radiation is absorbed by air, rock, and water, but is eventually reradiated into space at lower temperatures and longer wavelengths. The Earth has not become warmer through geologic time, so it must be assumed that as much heat is annually radiated away from the Earth as is received from the Sun and released from the interior.

The energy of the Sun is radiated over a very wide span of wavelengths in the electromagnetic spectrum. Solar flares emit very short X-rays; outbursts of solar "storms" also may emit very long waves that interfere with radio transmission on Earth. However, most of the solar energy is emitted in a narrow band of frequencies. The energy peak of the solar spectrum is in the wavelength of green light, near the center of the visible spectrum. Toward the shortwave, or ultraviolet, range of the spectrum, solar radiation drops off abruptly to less than 1/100,000 as much energy per micron of wavelength as at the peak. Toward the longwave, or infrared, range of the spectrum, the energy drops off somewhat more gradually, so that of the total radiant energy received, most is in the infrared portion of the spectrum, even though the peak wavelength is in the visible range.

The atmosphere acts as a selective filter to further restrict the wavelengths of solar radiation that reach the Earth's surface. Three constituents of the atmosphere are particularly efficient filters: ozone, water vapor, and carbon dioxide. Ozone (O_3) strongly absorbs most of the incoming ultraviolet radiation. Most of the ozone absorption takes place at an altitude of about 50 kilometers above the Earth, and because of the energy absorbed, the temperature climbs to nearly 0° C at that level of the atmosphere, in contrast to the frigid −60° C at both higher and lower levels. In the lower atmosphere, water vapor (H_2O) and carbon dioxide (CO_2) strongly absorb incoming solar radiation in selected longer infrared wavelengths. As a result, the radiant energy that finally penetrates the atmosphere is mainly in the near-visible infrared and visible wavelengths. It is one of the remarkable balances of nature that the Earth has been warmed by the Sun to just the right temperature so that it reradiates energy out through the atmosphere into space at a wavelength that passes through the atmosphere in a spectral "window" between the wavelengths strongly absorbed by water and by carbon dioxide. The steady average Earth temperature, then, is a result of balance between the wavelength and the intensity of both incoming and outgoing radiation.

The solar energy that penetrates the atmosphere falls mainly on the ocean. A small percentage of that energy is reflected by the water surface, and the remaining infrared energy is quickly absorbed in the upper few millimeters of sea water and converted into heat. The visible light, especially the bluish-green color, penetrates deeper into the ocean, for liquid water as well as atmospheric water vapor is highly transparent to visible light. Eventually, all of the light energy entering the sea is absorbed and converted to heat. Since the sea is efficient in transmitting heat by both conduction and convection, a layer as much as 100 meters in thickness is heated by the sun. This efficient mixing plus the high specific heat of water (the ability to store heat energy with only a slight temperature increase) make the oceans, especially the tropical oceans, great "heat exchangers" wherein solar energy is received and redistributed as heat.

Solar energy that reaches rocky surfaces and is not reflected is less efficiently absorbed. Rock-forming minerals are nearly opaque to visible light, except when ground and polished to very thin slices for microscopic examination, so the conversion from solar radiant energy to heat takes place at the surface of a rock. Although rock is a better thermal conductor than water, as a solid it cannot convectively circulate heat the way water does. Therefore, heat is scarcely transmitted into rock, but is concentrated near the surface.

Rock also has a lower specific heat than water. A small amount of heat raises the temperature of a rock much more than it will an equal mass of water. Recall the relief you have felt after walking barefoot across a blistering hot pavement or a sandy beach in bright sunlight to stand in the shallow water of a pool or ocean. Under the same bath of solar radiation, the water feels blissfully cooler than the rock or sand. As a temporary relief for your burning feet before you reached the water, you might have pushed them into the sand a few inches. The sand at even a slight depth is cooler than the surface layer, which absorbs almost all the energy.

Land reflects a higher percentage of the solar energy that falls on it than does water. Whereas the sea surface may reflect only 2 per cent of the incoming radiation, bare ground may reflect 7 to 20 per cent. Field crops or grasslands commonly reflect 20 to 25 per cent of the sunlight, but forests reflect only 3 to 10 per cent. From an airplane, fields look "brighter" than forests, and smooth water surfaces look darkest of all, for they reflect the least light to the eyes. Obviously, snow and ice are the most reflective surfaces. From half to nearly all of the sunshine may be reflected from an ice surface. All surfaces reflect more and absorb less energy when light strikes them at a low or grazing angle.

For several reasons, most of the solar energy is received by the Earth within the tropics. First, the Sun is high in the tropical sky at all seasons, and the reflectance is least when the incoming radiation is perpendicular to the surface. Second, the tropics contain large oceanic areas, favoring absorption of heat by water. Third, the surface area between successive parallels of latitude is larger near the equator than near the poles. Slightly more than a third of the entire surface of the globe lies between 20° north and 20° south of the equator. In sum, in the lower latitudes more energy is received from the Sun than is reradiated back into space.

The polar regions receive very little of the total solar radiation. The low angle of the Sun, even during the time of 24-hour daylight in high latitudes, promotes reflection. Both the Arctic Ocean and the Antarctic Continent are covered with ice, which further aids in reflecting the incoming radiation. It is true that for more than a month of midsummer, when the Sun shines continuously, each polar region in turn receives more radiation per day than is ever received per day in the tropics, but the season is very brief, and for the rest of the year the energy input at the poles is far lower than the output. Figure 1–2 summarizes the latitudinal distribution of incoming and outgoing radiation in the northern hemisphere, and shows that at a latitude of about 38° north, the annual heat budget of the Earth is balanced. The southern

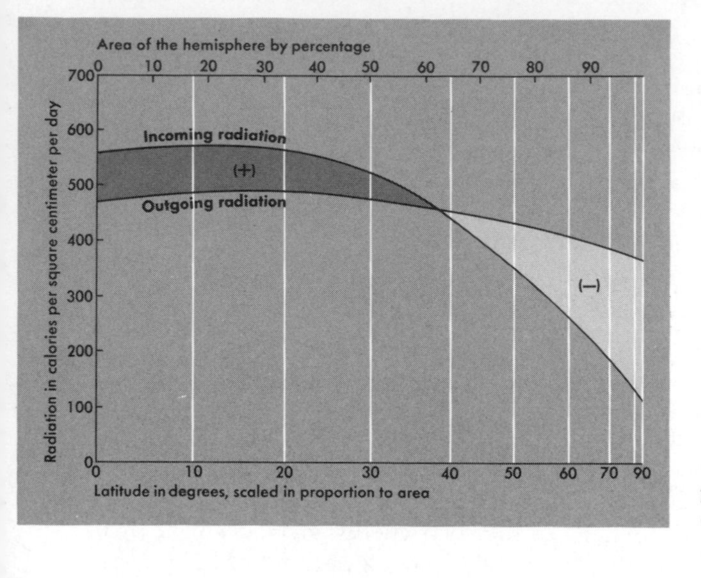

FIGURE 1–2 *Radiation balance of the northern hemisphere. Incoming radiation is the mean annual radiation absorbed by land, ocean, and atmosphere. Outgoing radiation is the annual amount that leaves the atmosphere. Within 38° north of the equator, more energy is received than is lost; north of 38° the Earth radiates more energy into space than it receives from the Sun. Heat energy must flow toward the pole. Presumably the southern hemisphere has a similar balance. (After Houghton, 1954.)*

hemisphere is assumed to have a similar budget. It is obvious that there must be ways for the Earth to distribute the excess incoming energy of the tropical regions toward the poles, for like any budget, this one must balance. Let us see how the energy distribution is accomplished.

The Hydrologic Cycle

It is a fortunate (or necessary?) condition that the Earth's average temperature is maintained in the range at which water remains liquid. Some ice and snow accumulate on the land at high altitudes and latitudes, and the atmosphere holds some water in the vapor phase, but nearly all of the H_2O of the Earth is liquid. The high specific heat of water and its great abundance near the surface of the Earth make it an effective agency of heat transfer. An even more important property of water is its very high *latent heat of vaporization.* Without changing the temperature of the water mass, 585 calories are required to evaporate a gram of water at 20°C (68°F). Conversely, each gram of rain that condenses and falls releases a similar amount of heat to the atmosphere at the point of condensation. Enormous amounts of energy are thus transferred when water is evaporated, carried elsewhere by winds, and condensed as rain or snow. Indeed, the largest single factor in heating the atmosphere is neither absorbed solar radiation nor reflected heat from the land and sea; it is the latent heat of vaporization released by condensation. Great quantities of water are annually evaporated in the tropical oceans and precipitated over lands and polar regions as rain and snow, in the process of maintaining the balanced heat budget of the Earth.

The primary result of the unequal heat and water distribution on Earth is to create climatic regions characterized by certain seasonal temperature and precipitation patterns. Climates are not the major topic of this book, although arid and glacial regions are given extensive special consideration.* The evaporation and precipitation of water are important to geomorphology, though, because it is this energy transfer that powers the geomorphology machine (Fig. 1–1). We can now be more quantitative and scientific about the geomorphology machine, and represent it by another kind of budgetary diagram (Fig. 1–3).

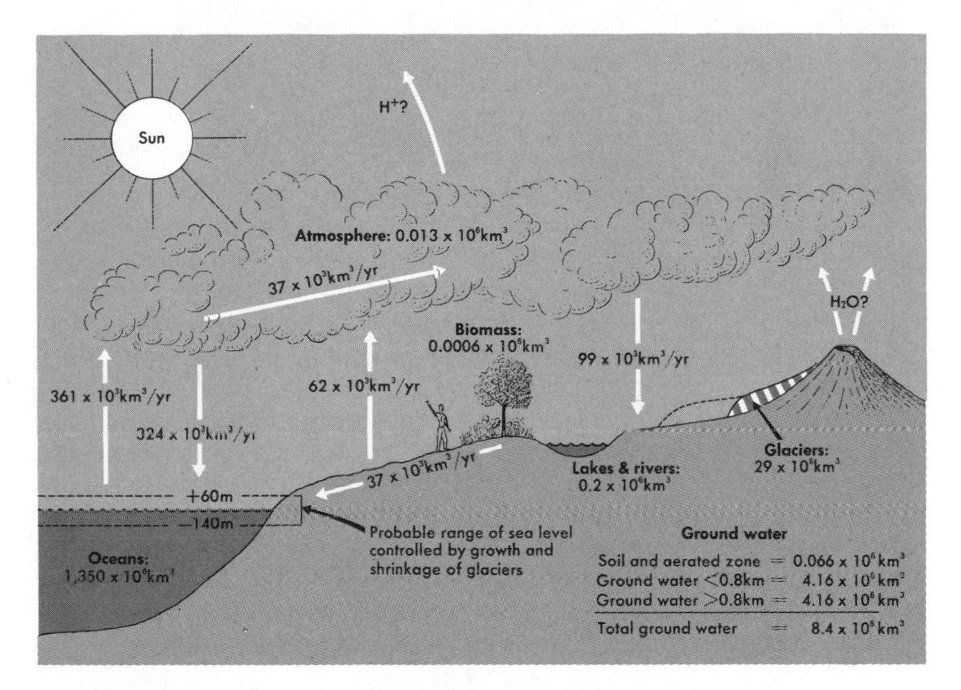

FIGURE 1–3 *The hydrosphere, or the water on or near the surface of the Earth. For easy comparisons, the amounts of water in the reservoirs are given in millions of cubic kilometers (10^6 km^3); the annual exchanges of water among the reservoirs, or the* hydrologic cycle, *are given in thousands of cubic kilometers per year (10^3 km^3/yr).*

This diagram should be examined with care, for it will be referred to many times in later chapters. Two kinds of information are summarized by Fig. 1–3: The first is an estimate of the *amounts* of water that are present in various parts of the outer layers of the Earth, the second is the calculated *annual exchange* of water between the various reservoirs. The figure portrays what we call a *steady-state system.* The energy inflow equals the outflow, and, with negligible exceptions, the amount of water in the system is constant.

*Another book in this series, *The Atmosphere*, gives fuller coverage to weather and climates.

First, examine the amounts of water in the system. In their order of importance, the reservoirs of water are: (1) oceans, (2) glaciers, (3) ground water, (4) lakes and rivers, (5) atmosphere, and (6) biomass (all living matter). Actually, over 97 per cent of all the water near the surface of the Earth is in the oceans, and most of the remainder is in glaciers. The interior of the Earth must contain some water, chemically combined or dissolved in solid or molten rock, but we cannot yet estimate how much water may still be locked within the Earth. It does seem though, that for at least the last billion years, the amount of water at or near the surface of the Earth has been approximately constant.

A small amount of new water might be added annually by condensation of volcanic gases, of which water is the principal component. Much steam issues from volcanoes, but almost all of it is either former rain water that has saturated the uppermost layers of rock, or former sea water that was trapped in marine sediments at the time of their deposition. The actual amount of new water annually added to the surface of the Earth by volcanic eruption, hot springs, and geysers has yet to be measured, but it is certainly very small.

An equally small amount of water is probably lost from the Earth annually by the photochemical dissociation of water vapor in the atmosphere by solar radiation. Some water molecules in the upper atmosphere are broken into H^+ and $O^=$ ions, and it is theorized that the velocity of the dissociated hydrogen ions is high enough for some of them to escape the gravitational field of the Earth. The annual amount of loss is very small. For all practical purposes, the total amount of water shown in Fig. 1–3 is constant, even over long geologic time. In terms of material transfer, the diagram shows a *closed system.* Of course, energy flows into, through, and out of the system.

The transfer of water between various surface and near-surface reservoirs is called the *hydrologic cycle.* It is usually shown as a rate of transfer, commonly as cubic kilometers or cubic miles of water per year. It can be converted into energy or power units, too. The quantities of water shown in Fig. 1–3 are mostly from recent estimates of hydrologists in the Water Resources Division of the U.S. Geological Survey. The values vary slightly from older estimates, but no high degree of accuracy is claimed for any of them.

We see from Fig. 1–3 that 361,000 km^3 of water are evaporated from the sea annually, representing a layer one meter thick from the entire ocean. Only 324,000 km^3 fall back into the sea, so a net surplus of 37,000 km^3 must go elsewhere. A smaller amount of water, 62,000 km^3, is annually evaporated from the land. This amount is actually smaller than the annual total precipitation on land (and ice) of 99,000 km^3. The net annual total excess of oceanic evaporation, then, falls on the land. And because more snow and rain fall every year than can be evaporated from the land, 37,000 km^3 of water annually drips, seeps, and flows from the land back into the sea.

Much of the water evaporated from the land is not simply evaporated from the exposed surfaces of lakes and streams, but is used by plants and animals. A field of crops may annually use an amount of water equivalent to a layer 18 to 24 inches deep over the field. Trees use even more water. A forest of Douglas fir, for instance, may annually pump into the atmosphere the equivalent of a layer of water four feet deep over its area. Over most of the land, rainwater falls on vegetation-covered ground. Some evaporates from the ground, some escapes from the leaves and stalks of plants, and some is absorbed by the plant roots and *transpired* through leaves. Since the several variables cannot be clearly distinguished, the term *evapotranspiration* is used for the collective effect. The water requirements of typical plants on the parts of the land that are neither desert nor glacier could evapotranspire all of the 62,000 cubic kilometers of water that annually evaporate from the land, so it is likely that each year a significant amount of water passes through a biologic cycle as a part of the larger hydrologic cycle. Only a small part of the evapotranspired water, perhaps 1 per cent, is held at any time as living matter, and the turnover is continuous.

Glaciers store large amounts of water on land, temporarily removing it from the hydrologic cycle. If the present glaciers were to melt, sea level would rise about 60 meters and submerge the most heavily populated areas of the Earth! In the last two million or so years, continental ice caps have repeatedly developed and melted again, each time temporarily upsetting the hydrologic cycle. During an age of maximum glaciation (see Chapter 7), the amount of glacier ice on land may have been three times as great as the present ice volume, and sea level may have been lowered as much as 140 meters, exposing most of the continental shelves.

All of these aspects of the hydrologic cycle relate to geomorphology, but one aspect in particular must be emphasized. The average continental height is 823 meters above sea level. If we assume that the 37,000 cubic kilometers of annual runoff flow downhill an average of 823 meters, the potential mechanical power of the system can be calculated. Potentially, the runoff from all lands would continuously generate over 12 billion horsepower. If all this power were used to erode the land, it would be comparable to having one horse-drawn scraper or scoop at work on each three-acre piece of land, day and night, year around. Imagine the work that would be accomplished! Of course, a large part of the potential energy of the runoff is wasted as frictional heat by the turbulent flow and splashing of water, but we will see that the "geomorphology machine" is really quite efficient, and in fact does erode and transport rock debris down to the sea almost as fast as if horse-drawn scrapers were hard at work on every small plot of land, over all the Earth.

2

Rock weathering

The physical and chemical alteration of rock when it is exposed to the atmosphere is called *rock weathering*. The term is a good one, recalling the common expression "weather-beaten" that is used to describe buildings and even human faces that have been marked by exposure to sun, wind, and rain. A fascinating variety of landscapes can result from weathering, and a trained observer can make a good guess about the rock types and climate of a region just from a glance at the weathered rock debris he sees from a car window.

Weathering can be considered as a process of metamorphism. We usually think of metamorphism as the alteration of rocks in response to increased temperature and pressure, but rocks also change their properties when the temperature and pressure of the environment are decreased. Most rocks are formed under temperatures and pressures that are higher than the average conditions at the surface of the Earth. In general, then, rocks are both physically and chemically unstable when exposed to a wet and biologically active atmosphere under a pressure of only 14.7 pounds per square inch, in the temperature range of about 0°C to 100°C.

Basically, if artificially, weathering processes are subdivided into a mechanical, or physical, group and a chemical group. Mechanical weathering is sometimes called *disintegration*, implying that pieces of rock are taken apart or segregated without alteration. In contrast, chemical weathering is sometimes called *decomposition*, emphasizing the breakdown of the chemical structure in the mineral grains that make a rock. The distinction is not as important as it formerly was, for as we approach an understanding of the two groups of processes we find that the same principles govern both. It will be convenient, though, to consider mechanical weathering processes first, because large-scale mechanical fracturing of rocks usually is necessary before air, water, and organisms can begin their largely chemical attack.

Mechanical Weathering

The most important processes by which rocks are mechanically broken or disintegrated are: (1) differential expansion by pressure release when the rock is exposed at the surface, (2) growth of foreign crystals such as ice or salt in cracks and pores, and (3) differential expansion and contraction during unequal or rapid heating and cooling. Each of these processes affects different rock types in different ways, and the second and third processes are strongly dependent on climatic conditions.

Pressure Release on Unloading

The atmospheric pressure at the surface of the Earth is much less than the pressures encountered even at shallow depths within the Earth or beneath the sea. Each 10 meters of sea water or three to four meters of overlying rock are equivalent to another atmosphere of confining pressure. Whether a rock was formed by consolidation of mud at the bottom of the sea or by cooling of a molten mass deep within the Earth, it was subjected to greater pressure when it formed than when it is exposed to the atmosphere. Uplift of the land has drained away the sea and erosion has removed hundreds or thousands of feet of overlying rock to expose the particular rocks that now form our landscape. When so unloaded, rocks expand from the pressure release.

During the time that they are forming within the Earth, or during the uplift that must precede their eventual exposure by erosion, most rocks are thoroughly fractured into blocks or slabs that range in length from a few inches to several feet. The fractures, or *joints*, commonly have systematic regional orientation and spacing that can be related to the forces that caused the uplift (Fig. 2–1). If a rock is well jointed, or if a sedimentary rock has open spaces along *bedding planes* (original sedimentary layers), the slight expansion caused by unloading is relieved by movement on these pre-existing surfaces. How-

FIGURE 2–1 *Jointed sandstone and shale near Ithaca, New York. Sedimentary bedding is obvious. The joints belong to two vertical sets, one set trending northeast and the other northwest. Other oblique joints are less well developed.*

ever, some rocks, especially sandstone and granite, are massive, with few open joints or other inherent fractures. A distinctive set of joints, called *sheeting joints* or simply *sheeting*, develops in these massive rocks when they are exposed by erosion. Sheeting joints parallel the land surface and form concentric shells or layers of rock up to a few feet in thickness (Fig. 2–2). They form progressively downward as the outermost shells of the rock are relieved of load and expand, thus breaking away from the rock beneath.

The amount of locked-in expansive force in massive rocks is surprising. For instance, newly quarried blocks of granite from Stone Mountain, Georgia (Fig. 2–2) expand one-tenth of one per cent along their length when they are

FIGURE 2–2 *Sheeting joints on the southeast flank of Stone Mountain, a granite dome near Atlanta, Georgia. Note the close parallelism of the sheeting with the surface of the dome. (Courtesy C. A. Hopson.)*

cut free of a quarry wall. A ten-foot block of this granite will expand almost one-eighth of an inch when it is cut or blasted loose. No wonder that when thin surface slabs of rock break loose, they arch upward from the still confined rock beneath.

Sheeting joints rarely extend more than a few hundred feet below the land surface. The weight of the overlying rock prevents expansion at greater depths. In mines and quarries that penetrate deeply into massive rocks, special precautions must be taken to prevent "rock bursts" or the explosive expansion of slabs from the walls or floors of newly opened working spaces. Blasting operations are usually scheduled at the end of the working day, so that the rock faces can adjust to the new pressure environment overnight without endangering workmen or mining equipment.

Pressure release by unloading is only a minor factor of weathering by itself, but it is often the first event in a continuing series. Unless water, air, and plant roots can penetrate massive rock along sheeting joints and similar fractures, other kinds of weathering are prevented.

Growth of Foreign Crystals within a Rock

If water is confined in a crack in a rock and freezes, the expansion on freezing generates very large stresses within the rock. When water freezes under atmospheric conditions, its molecules organize into a rigid hexagonal crystalline network and it increases 9 per cent in specific volume (volume per unit mass). Intuition will tell you that a transformation of this sort, which produces a volume increase, will be inhibited by confining pressure. The higher the confining pressure, the more the temperature must be lowered in order to freeze water.

A very high confining pressure is required to depress the freezing point of water only a few degrees. For instance, a confining pressure equal to 150 atmospheres, or approximately 150 tons per square foot, will lower the freezing temperature of water only 2°F. Conversely, if confinement prevents water from beginning to freeze until it is cooled to a temperature of 30°F, a pressure of 150 tons per square foot is generated against the walls of the container. A graph of the relation between the freezing temperature of water and confining pressure is shown in Fig. 2–3.

Water in shallow soil, or standing on the surface of the ground, freezes at 0°C (32°F) and freely expands 9 per cent because it is unconfined. Small pebbles, garden soil, and grassy turf are "heaved" by unconfined freezing, but no great pressure is involved. When the water filling a crack in a rock begins to freeze, however, the surface layer freezes first and confines the water beneath. As more ice freezes and expands, the pressure on the remaining water increases and the temperature must be lowered further in order to continue the process. Pressure continues to build up (Fig. 2–3) until the crucial temperature of − 22°C (−7.6°F) is reached. At that temperature, confined water

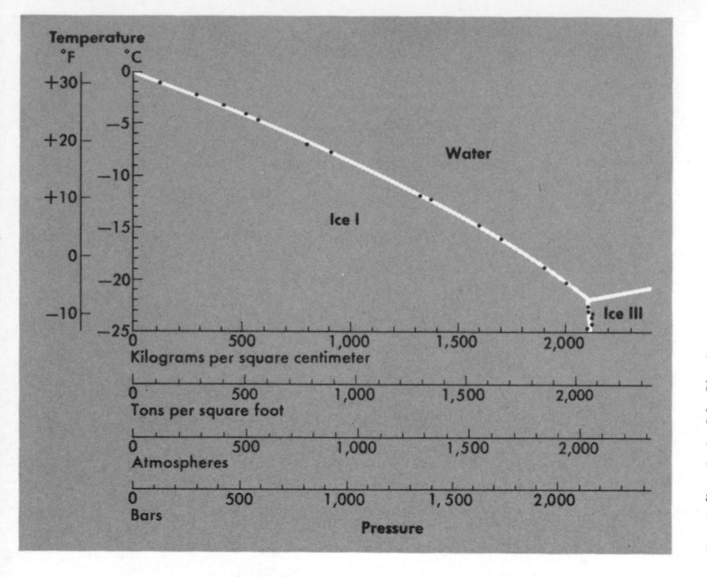

FIGURE 2–3 *The pressure increase in confined pure water with lowered freezing temperature. At confining pressures greater than about 2,100 tons per square foot, more dense forms of ice begin to crystallize, one of which, Ice III, is shown. (Data from Bridgman, 1911.)*

exerts a maximum pressure of almost 2,100 tons per square foot. Visualize this as a navy destroyer standing on a pedestal one foot square. All but the strongest rocks are crushed by such pressure. At freezing temperatures below −22°C, the pressure remains about constant, because a more dense kind of ice begins to crystallize. (Experimental tests in huge refrigerated presses have verified at least seven distinct crystalline forms of ice, but only the well-known hexagonal form is found in nature.)

Ice is not strong enough to seal water into a rock crack and produce the maximum possible pressure. Instead, the ice is pushed out of the crack as a plug. Furthermore, water in very thin cracks will not freeze, even at very low temperatures, because of the strong capillary adhesion of the water to rock. It has been claimed that thin capillary films of water develop a semi-crystalline molecular structure on rock surfaces, and disrupt the rock by expansion even without freezing. Studies of such "ordered films" of water emphasize the disappearing distinction between mechanical and chemical weathering processes.

The term *frost wedging* describes the process by which freezing water breaks rocks apart. Even if the maximum pressure of 2,100 tons per square foot is not realized in nature, substantial forces are generated when water freezes in jointed or thinly bedded rock. Frost wedging is most effective in humid climates where the daily temperature ranges through the freezing point. In such a climate, as for instance on a sunny but cold mountain top, meltwater saturates the rocks during each day and freezes each night. Many mountain slopes above the tree line are wastelands of frost-shattered rock rubble that is gradually being broken down to particle sizes that can be attacked by other weathering and erosional agents.

Ice is not the only crystalline material that can accumulate in cracked rocks and pry the cracks open. Many salts accumulate in joints and even along the boundaries of mineral grains when water-saturated rocks are dried. Tests were conducted by the U.S. National Bureau of Standards to determine the causes for the scaling and crumbling of granite buildings and monuments. It was found that blocks of construction-grade granite could be subjected to 5,000 cycles of alternate freezing (six hours at −12°C) and thawing (one hour in water at 20°C) with barely visible disintegration, but when blocks of the same granite were alternately soaked in a saturated solution of sodium sulfate (17 hours at room temperature) and dried (seven hours at 105°C), they crumbled after an average of only 42 cycles. Crystals of water-soluble salts precipitate from solutions that are supersaturated by evaporation, and as they grow they quickly disintegrate any rock that water can penetrate.

We think of rain water as being pure, but rain always contains traces of dissolved substances, and after water has flowed even a short distance over rock or through soil, it contains even more dissolved salts that can later precipitate in rock openings. A particularly troublesome salt is hydrous calcium sulfate (the mineral *gypsum*). In cities where the air is polluted by sulfurous coal smoke, rainwater becomes a dilute sulfuric acid that violently corrodes limestone and marble buildings. The reaction product of the corrosion is gypsum, a relatively insoluble salt that crystallizes in rock cracks, flakes off thin pieces of rock, and accelerates chemical attack. Many ancient European monuments have been severely damaged by the combined attack of smoky, moist air and precipitated gypsum. The formation of salt crystals in rock openings is regarded as a mechanical weathering process, but the close relationship with chemical processes is obvious.

Thermal Expansion and Contraction

Boulders and cobbles exposed to hot desert sun are often found broken in sections somewhat like orange slices. Desert travelers have described hearing the rifle-shot sounds of stones breaking, and French Foreign Legion forts in the Sahara are said to have gone to battle alert at the sounds. Thermal expansion and contraction, due to hot days and rapid cooling during the desert evening, have been claimed as the cause of the cracked stones.

Strangely, no one has ever verified experimentally that solar heating is intense enough to break stones. Specimens of the same granites that were subjected to the freeze-and-thaw and salt-solution tests of the U.S. Bureau of Standards just described were heated dry to 105°C and cooled to −10°C for 2,000 cycles with no evidence of deterioration. In another often-quoted study published in 1936, D. T. Griggs alternately heated the surface of a three-inch cube of dry granite for five minutes to about 140°C, then cooled it for ten minutes with an electric fan to about 30°C, for 89,400 cycles, equal to 244 years of diurnal heating. His experiment lasted over three years, but no change in the rock could be detected, even by microscopic examination. His experiment is usually cited as evidence that solar heating and atmospheric cooling are not sufficient to crack rocks. Yet the reports come in, most recently from

Australia, that many types of rocks are found shattered in the desert with no known weathering process except temperature change as the likely cause. This is one of the intriguing minor problems of geomorphology. Good experimental design and patience should be sufficient to solve it. Incidentally, even if the process is insignificant on Earth, it is likely to be one of the major processes of lunar "weathering."

Plants as Agents of Mechanical Weathering

The prying or wedging action of plant roots, especially of trees, is often described as mechanical weathering. Two-dimensional networks or sheets of interlaced roots can be followed for many feet along bedding planes or joints, deep into fresh rock. It has been supposed that the growing roots exert a pressure on the rock and force cracks to open. The efficacy of roots as agents of mechanical weathering probably has been overestimated, but their importance as agents of chemical weathering probably has been underestimated. Roots follow the paths of least resistance, and conform to each little irregularity of a crack, but they do not seem to exert much force on the rock. Cracks opened by other processes can be maintained by roots, however, and decaying vegetable matter and washed-in dirt can keep rock surfaces wet and chemically active. In addition, of course, when trees sway in a strong wind their roots powerfully pry apart rocks. Forest soils usually include many mounds and pits formed by tree roots that have been torn up when trees fell.

Chemical Weathering

The general trends of chemical weathering can be predicted from the conditions under which rocks form. Having formed in an environment of high thermal energy and high pressure, rocks at the surface of the Earth tend to weather by exothermic (heat-evolving) chemical reactions that produce new compounds of greater volume and lower density. *Oxidation* is one of the most typical exothermic, volume-increasing reactions between rocks and the wet atmosphere; especially common is the reaction of iron-bearing minerals with oxygen dissolved in water. Other typical weathering reactions are *carbonation*, the reaction of minerals with dissolved CO_2 in water; *hydrolysis*, or decomposition and reaction with water; *hydration*, the addition of water to the molecular structure of a mineral; *base-exchange*, the exchange of one cation (positively charged ion) for another between a solution and a mineral solid; and *chelation*, the incorporation of cations from the mineral into organic compounds.

All weathering reactions involve water either as a reactant or as the carrier of the reaction products. It is not necessary, therefore, to list *solution* as a unique chemical weathering process, for solutions in water are always a part of chemical weathering. Water is so important to rock weathering that we must consider some of its properties further before we examine the specific chemical weathering reactions.

Water in Chemical Weathering

Water is a peculiar compound, in spite of our everyday acceptance of it. We must recognize some of its odd chemical and physical properties to appreciate rock weathering. First, water is by far the most abundant material near the surface of the Earth. In the outer five kilometers of the Earth, water is about three times as abundant as all other substances together, and about six times as abundant as the next most common single substance, the mineral feldspar. Water is the only compound that occurs naturally at the Earth's surface in gaseous, liquid, and solid states. Its general solvent power and its surface tension are greater than those of any other fluid. Its heat of vaporization is the highest of all substances; recall the significance of this property in the conversion of solar energy to geomorphic work. Its maximum density at 4°C in the liquid phase, with expansion toward the freezing temperature as well as away from it, is a strange and almost unique property. The 9 percent expansion of water on freezing is exceptionally great. A recitation of the properties of water could go on at length, but this summary is enough to suggest that the surface of the Earth is generally saturated with an abundant and chemically active compound, made available in unending supply by the hydrologic cycle, that readily attacks and reacts with rock-forming minerals.

An interesting way to demonstrate the rapid reaction of water with rock is to trace, by means of successive chemical analyses, the changes in dissolved compounds in water after it falls on the land as rain or snow. A particularly thorough study of this sort was published in 1964 by J. H. Feth and others of the U.S. Geologic Survey, Water Resources Division. Working in the Sierra Nevada, they analyzed new-fallen snow, meltwater from the base of snowbanks, water that had soaked into the soil at the base of snowbanks, and water that had penetrated deeply into joints and other rock openings and reappeared at the surface from permanent springs.

The amount of dissolved matter in the snow was only a few parts per million, mostly sodium, chloride, and bicarbonate derived from sea spray and atmospheric carbon dioxide. Other components were dust and gases from a variety of sources. Each snowfall had a different composition, depending on the history of the air mass that carried it. As soon as the snow meltwater soaked into the mountain soil, the mineral content of the water increased seven and one-half times. The greatest increase in dissolved compounds was shown by silica, which increased 100-fold almost as soon as the water entered the ground (but nevertheless formed only a small fraction of the total dissolved material). Water that moved deeper underground for several months, forming permanently flowing springs through the dry season, doubled its mineral content as compared to the water that had only briefly soaked into the ground. A striking conclusion of the study is that about half of the mineral content of the ground water was acquired during the first few hours to few weeks of contact between melting snow and soil.

We have seen that the rainwater that is distilled from the salty sea by sunshine and falls on the land has potential mechanical energy by virtue of the product of its mass and altitude above sealevel; it also has great potential chemical energy by virtue of its great chemical contrast to the minerals of the rocky Earth. Whereas the potential energy due to altitude is gradually converted to kinetic energy as a raindrop joins the water of a river and returns to the sea, the chemical potential of a raindrop is apparently largely expended by the various reactions with mineral matter soon after the drop contacts the ground. In general, river water has about the dissolved chemical load that could be predicted from analyzing the rocks through and over which it has flowed.

Oxidation

Weathering by oxidation probably always takes place with water as the intermediary. Unprotected iron surfaces stay bright and clean in dry air, but they rust, or oxidize, quickly in the presence of moisture. There is always enough dissolved oxygen in rainwater and circulating groundwater to oxidize metallic iron, and to change the ferrous iron in mineral compounds to the more oxidized ferric state. As long as water is in contact with atmospheric molecular oxygen on the one hand and incompletely oxidized iron on the other, oxygen dissolves from the air, diffuses through the water, and combines with the iron. When the iron or other elements from a mineral grain combine with oxygen, the original mineral structure is destroyed, and the remaining mineral components are free to participate in other chemical reactions.

Weathering by oxidation occurs on exposed rock surfaces, and is typically indicated by a red or yellow surface layer on the weathered rock. Many kinds of rocks contain traces of iron, which is a very common chemical element, and regardless of their natural color, all weather to a similar "rusty" yellow or red-brown. For this reason, always break open a rock and examine a freshly exposed surface if you wish to describe the color of the rock. Oxidation also affects elements other than iron, of course, but because of its abundance and easy oxidation, iron is ideal to show the general character of this kind of weathering.

Iron oxides are exceptionally stable chemical compounds, and once formed, they are destroyed only by chemical reduction with carbon. Most iron ores are mixtures of iron oxides and mineral impurities. The smelting of iron ore requires carbon, from coal, coke, or petroleum, and also a high temperature for the reaction. When we make iron or steel from an oxide-rich iron ore, we must expend a large amount of heat energy to reverse a weathering process.

Some organic processes can reduce iron oxide from the highly oxidized ferric state to the less oxidized ferrous state. Ferrous iron compounds are more soluble in water than ferric compounds, so reduction by organisms is one way that iron compounds can be removed from weathered rock or soil. The anaerobic bacteria, a group of organisms that can live without free oxygen, even reduce iron oxide completely to metallic iron, in order to use the derived oxygen for their metabolic processes. Such reducing "weathering" processes, which take place in stagnant water or foul-smelling, organic-rich mud, are not typical chemical reactions under Earthly atmospheric conditions. Red, yellow, or brown colors in sediments or sedimentary rocks almost always indicate deposition in a well-oxidized environment. The environment need not have been exposed to the air, however, for circulating ocean and lake water is well supplied with dissolved oxygen.

Carbonation

Carbon dioxide gas dissolves readily in water. Cold water can dissolve more carbon dioxide gas than warm, and water under increased pressure can also dissolve more gas. Dissolved carbon dioxide combines with water to form a weak acid (carbonic acid) as shown by the following reaction:

$$CO_2 + H_2O \rightarrow H_2CO_3$$

gas + water = carbonic acid

Fresh rainwater is always slightly acid because of the carbon dioxide it has dissolved from the atmosphere. The acidity increases further when the water soaks through rotting vegetation in the upper layers of soil, for the air trapped in soil is unusually rich in carbon dioxide generated by decaying plant tissue. The biologically generated carbon dioxide in soil air is the major source of carbonated ground water.

The reaction between carbonic acid and minerals is called *carbonation.* Because all water in contact with air contains at least some dissolved carbon dioxide, carbonation is a common process of weathering.

Carbonation is most obvious in the weathering of carbonate rocks such as limestone. Calcium carbonate, as the common mineral calcite, is the principal compound in limestone. It is not very soluble in pure water, but in the presence of carbonic acid, the following reaction takes place:

$$CaCO_3 + H_2CO_3 \rightarrow Ca^{++} + 2\ HCO_3^-$$

calcite + carbonic acid = calcium and bicarbonate ions in solution

Calcium bicarbonate is about 30 times more soluble than calcium carbonate in water, hence the reaction causes rapid dissolution of limestone.

Most limestone contains some insoluble impurities, such as clay and quartz sand, that accumulate to form a residual soil as the limestone dissolves. The residual iron-bearing minerals are commonly oxidized and therefore red, and one of the striking scenic features of many limestone regions is the red clay soil overlying gray or white limestone. Depending on the proportion of soluble to insoluble minerals, as much as 10 to 20 feet of limestone must be weathered away to produce a residual soil a few feet thick. Therefore, the soil on limestone represents a considerable lowering of the land surface. It is estimated that in the limestone region around Mammoth Cave, Kentucky, the landscape is lowered one foot every 2,000 years by carbonate solution. In the chalk hills of England, solution by carbonated water has lowered the land surface 15 to 20 inches in 4,000 years. This interesting measurement was made by comparing the height of the bedrock floor beneath prehistoric earthworks with the rock surface beneath the surrounding soil. The clayey soil that was used to build the mounds has shed rainwater and preserved the underlying rock surface, while the surrounding landscape has been continuously lowered by solution. The age of the mounds can be closely estimated by the artifacts found in them.

Limestone solution not only lowers land surfaces, but it may also operate at depth. Limestone caverns form where subterranean water circulates, initially along joints or bedding planes and subsequently in conduits formed by solution of the limestone. An area underlain by limestone can be weathered into a spongy network of caves and tubes. Limestone caves may extend hundreds or even thousands of feet below the land surface. A landscape displaying solution features on limestone is called a *karst landscape*, after the characteristic Karst Region of Yugoslavia. Some of the special landforms of karst regions are described later in this chapter.

Carbonated water that has been in contact with limestone is "hard," or saturated with calcium bicarbonate. Either decrease in pressure, increase in temperature, or evaporation will cause supersaturation and precipitation of some limestone. The thick layer of lime in an old teakettle or water boiler is caused by all three changes. Stalactites, stalagmites, and other dripstone formations in caves (Fig. 2–4) are commonly regarded as the result of evaporation only. However, cave air is highly humid and evaporation is slight. At least some of the deposition is caused when groundwater, moving under pressure through the rock above the cave, enters the free air of the cave and loses some CO_2 because of the drop in pressure. With the loss of carbon dioxide from solution, some dissolved calcium bicarbonate reverts to less soluble calcium carbonate, usually at the tip of a projecting ledge over which the water drips or flows. The resulting calcite deposits, called flowstone, dripstone, or travertine, build the weirdly contorted pillars, curtains, and terraces that make caves major tourist attractions.

The carbonation of silicate minerals, which form most rocks, is not as rapid or simple as the reaction of carbonic acid and limestone. However, carbonation is a significant weathering process in all rocks because it is a first step toward hydrolysis, which will be described next.

FIGURE 2–4 *Dripstone formations in the Postojna Caves, Yugoslavia. This is one of the most popular tourist caves in the famous Karst Region of Yugoslavia. (Courtesy Yugoslav State Tourist Office.)*

Hydrolysis

The most important chemical weathering reaction of silicate minerals is *hydrolysis*, or decomposition and reaction with water. In hydrolysis, water is not just a carrier of dissolved reactants, but is itself one of the reactants. Pure water ionizes only slightly, but it does react with some easily weathered silicate minerals in the following fashion:

$$Mg_2SiO_4 + 4H^+ + 4\,OH^- \rightarrow 2\,Mg^{++} + 4\,OH^- + H_4SiO_4$$

olivine + 4 ionized water molecules = ions in solution + silicic acid in solution

The result of such complete hydrolysis is that the mineral is entirely dissolved, assuming that a great excess of water is available to carry the ions in solution. Silicic acid, one of the reaction products, is such a weak acid that we can disregard its name and simply think of it as silica (SiO_2) dissolved in water.

Any reaction that increases the H^+ ion concentration in water also increases the effectiveness of hydrolysis. Carbon dioxide dissolving in water is the most common and important way that water is provided with H^+ ions, or acidified, for hydrolysis. We can elaborate the chemical equation previously given for the reaction between water and carbon dioxide as follows:

$$CO_2 + H_2O \rightleftharpoons H_2CO_3 \rightleftharpoons H^+ + HCO_3^-$$

gas + water = carbonic acid = hydrogen ion + bicarbonate ion

Each double arrow means that an equilibrium is established, with all compounds or ions present simultaneously in the solution. If any component is removed, the equilibrium will shift in a direction that tends to restore the loss. Specifically, during the hydrolysis of the silicate mineral olivine by carbonated water (paralleling the previous example of hydrolysis by pure water), as H^+ ions react with the olivine to form silicic acid, more carbonic acid ionizes to restore the balance, and, in turn, the carbonic acid concentration is maintained by more carbon dioxide dissolving into the water. Carbonic acid is a much better supplier of H^+ ion for hydrolysis than pure water would be, and the byproducts of the hydrolysis are soluble and easily carried away.

Probably the most common weathering reaction on Earth is the hydrolysis of feldspar minerals by carbonic acid. Feldspar, the family name for a large group of potassium, sodium, and calcium alumino-silicate minerals, is second only to water in abundance in the outer several miles of the Earth's crust. A typical, although simplified, weathering reaction between potassium feldspar (orthoclase) and carbonated water is as follows:

$$2\,KAlSi_3O_8 + 2\,H_2CO_3 + 9\,H_2O \rightarrow Al_2Si_2O_5(OH)_4 + 4\,H_4SiO_4 + 2\,K^+ + 2\,HCO_3^-$$

orthoclase + carbonic acid + water = kaolinite—a clay mineral + silicic acid in solution + potassium and bicarbonate ions in solution

Calcium and sodium feldspars (collectively called plagioclase) hydrolyze even more readily in acidic water than does orthoclase. All feldspar hydrolysis in carbonated water gives three end-products: (1) a clay mineral, (2) silica in solution, and (3) a carbonate or bicarbonate of potassium, sodium, or calcium in solution. The clay minerals are stable residual solids under all but the most humid and tropical climatic conditions. Much of the potassium freed by hydrolysis is subsequently either incorporated into clay minerals other than kaolinite, or used by plants (it is an essential plant nutrient). Therefore, river water and the sea have less potassium than you would expect from the abundance of weathered orthoclase. Sodium, calcium, and bicarbonate ions however, are common in river water. The sodium accumulates in the sea, but most of the calcium and bicarbonate ions that reach the sea are subsequently absorbed by marine organisms that use them to build their limey skeletal frameworks or shells. The silica in solution is also used by organisms, largely the single-celled plants called diatoms, and sea water is depleted in silica relative to the amount in rivers.

Hydration

Weathering by *hydration* involves the addition of the entire water molecule to the mineral structure. Water of hydration causes minerals to expand and is therefore considered by some scientists to be a process of physical weathering, related to the growth of foreign crystals in a rock. Water of hydration can be driven off by heating minerals above the boiling point of water. In contrast, water that reacts with feldspar during hydrolysis becomes part of the atomic structure of a clay mineral and can be driven off only by destroying the mineral at high temperature.

Many of the clay minerals are hydrated. The closely related hydration and hydrolysis of feldspars to form clays are the major processes in the weathering of granite. Weathering feldspar grains expand and cause the granite to crumble into a mass of disaggregated grains of quartz and weathered feldspar called *grus*. Sometimes weathered granite can be tunneled with ordinary picks and shovels, as many prospectors know.

Some clay minerals will hydrate and dehydrate alternately, depending on the moisture available. After a heavy rain, they expand and heave overlying soil or rocks. On drying, they shrink and crack. These "swelling clays" create major engineering problems where they are found.

Base-exchange and Chelation

Soil scientists in recent years have recognized the importance of two other weathering reactions, base-exchange and chelation. *Base-exchange* involves a mutual transfer of cations such as Ca^{++}, Mg^{++}, Na^{+}, or K^{+}, between an aqueous solution rich in one cation and a mineral rich in another. The rate of exchange depends on the chemical activity and the abundance of the various cations, as well as on the acidity, temperature, and other properties of the solution. The principle of base-exchange is used to improve soil fertility by adding solutions rich in the needed substance. It is also the principle by which "hard" water, usually rich in calcium bicarbonate, is "softened" by exchanging sodium ions for the calcium ions in the water. The exchange of cations between minerals and ground water may expand or collapse the mineral structure and free other chemical components. As in other chemical reactions, if one mineral grain in a rock is so destroyed, adjacent grains are detached and exposed to the same and other weathering processes.

Chelation is a complex organic process by which metallic cations are incorporated into hydrocarbon molecules. The word "chelate," which means "claw-like," refers to the tight chemical bonds that hydrocarbons may impose on metallic cations. Many organic processes, in order to function, require metallic-organic chelates. The role of iron in hemoglobin to carry oxygen from the lungs is a good example.

Plant rootlets maintain a field of charged H^+ ions around their tips that is strong enough to hydrolyze minerals and put vital metallic cations into solution so that the plant can chelate and absorb them. In the process of chelation, metabolic energy derived directly from sunlight falling on the leaves of a plant is used to weather minerals underground.

Weatherability of Silicate Minerals

By comparing the rates of chemical reactions between water and various powdered minerals in controlled laboratory experiments, it is possible to predict which minerals will weather most rapidly under natural conditions. Extensive studies of this sort, especially on the silicate minerals that form most of the Earth's crust, have led to the recognition of a *weathering series,* or a list of common silicate minerals arranged in order of relative susceptibility to chemical weathering. It is notable that the silicate minerals that crystallize first from a molten mixture, when the temperature is still very high, react the most rapidly with water in chemical weathering. Olivine, the mineral that was used to illustrate hydrolysis in pure water, is a typical high-temperature mineral that weathers easily at the Earth's surface. Calcium-rich plagioclase feldspar weathers easier than sodium-rich plagioclase, and potassium feldspar (orthoclase) is the least reactive member of the feldspar family. In a cooling silicate melt, calcium-, then sodium-, then potassium-rich feldspar crystallizes as the temperature falls, and the relationship between temperature of formation and ease of weathering is maintained. The last major mineral to crystallize from a silicate melt is quartz (SiO_2), and quartz is by far the most resistant rock-forming silicate mineral.* The silica in solution in ground water is not dissolved quartz but one of the end products of weathering of other silicate minerals.

* The sequential crystallization of minerals from a silicate melt is known as "Bowen's reaction series" and is discussed further in W. G. Ernst's book *Earth Materials,* in this series.

Igneous and metamorphic rocks commonly crystallize as granular masses. If quartz grains are present, they are freed from the rock when other minerals are decomposed. The quartz grains are typically sand sized. In fact, the reason for the abundance of quartz sand in river channels and on beaches is that so much of it is freed from rocks by chemical weathering of the surrounding mineral grains. Quartz is hard and resists abrasion. It is also virtually insoluble, so once liberated by weathering, quartz sand is carried to the sea, eventually to form sandstone that may be uplifted and exposed to the atmosphere, weather and disintegrate to sand again, and once more be transported to the sea to form a second-generation sandstone. Some sand grains have probably survived several such cycles of weathering, erosion, deposition, uplift, and renewed weathering.

Climates and Weathering

We have already seen, in examining frost wedging, oxidation, and other specific weathering processes, that climate plays an important role in determining the kinds and intensities of weathering processes. Climate controls weathering directly, through the temperature and precipitation of a region, and also indirectly, through the kinds of vegetation that can cover the landscape. We will consider four kinds of climates, noting the characteristic weathering processes and the nature of the resulting weathered materials.

Humid Tropics

The best examples of extreme chemical weathering are certain tropical soils that are such concentrated iron or aluminum oxides that they are widely mined as ores. Rock-hard soil rich in iron oxide is called *laterite,* and soil rich in aluminum oxide is called *bauxite.* The extreme weathering that these soils represent can be illustrated by the following reaction, which begins with kaolinite, the clay mineral that was shown as one of the end products of less extreme weathering of orthoclase in the previous example of hydrolysis:

$$\underset{\text{kaolinite—a clay mineral}}{Al_2Si_2O_5(OH)_4} + \underset{\text{water}}{5H_2O} \rightarrow \underset{\text{hydrated aluminum oxide}}{Al_2O_3 \cdot 3H_2O} + \underset{\text{silicic acid in solution}}{2H_4SiO_4}$$

The most favorable conditions for such complete decomposition of minerals are found in tropical climates with heavy annual rainfall and at least a brief dry season. The hydrated aluminum oxide eventually crystallizes as a component of bauxite, the earthy, impure crystalline material that is the principal ore of aluminum.

Iron-rich minerals undergo similar extreme tropical weathering, and the residual lateritic soils sometimes harden when exposed to air and can be used for construction materials. The ancient temples at Angkor Wat, Cambodia, are built in part of quarried laterite. Both laterite and bauxite can be regarded as the ashes of "burned out" tropical soils. All that is left are the very stable oxides of iron or aluminum.

The silica that is dissolved from tropical soils is either removed to the sea, or, if insufficient water is available to flush the silica from the soil, it may dehydrate to form amorphous silica crusts or layers in the weathered rock. *Opal* is a semiprecious gem stone that forms in weathered rock from dehydrated colloidal silica.

In the humid tropics, aluminum-rich clays characterize the weathered layer, which may be hundreds of feet deep. Strong red and yellow colors of oxidized iron stain the weathered rock, but silica and other soluble compounds may be nearly completely removed. Plants derive nutrients either by very deep root-penetration or by re-using the nutrients from dead plants. Dead plants and animals are immediately attacked by hosts of scavenging micro-organisms, and the nutrients are recycled through the food chain. Unfortunately for tropical agriculture, if even one crop is removed from the soil, the remaining nutrients may no longer support vegetation and the bare red earth may even harden permanently to brick-like laterite.

Limestone in the tropics is carbonated and dissolved intensely both by carbonic acid solutions and by acidic nitrogenous and organic compounds from the decaying vegetation. Tropical weathering of limestone may reduce a region to a series of sponge-like hills and cavern-ridden lowlands. Few streams flow across tropical karst landscapes, for most of the drainage is underground. The strange, needlelike mountains of classical Chinese art originated in the tropical karst landscape of the southern provinces of China (Fig. 2–5). The scenes seem exotic and dream-like to Western eyes, but they are actually reasonable geomorphic sketches.

Humid Mid-latitudes

In a humid climate with seasonal freezing, frost wedging assumes importance as a mechanical weathering process. Cold weather brings a dormant period for both vegetation and soil micro-organisms, so there are less severe demands for mineral nutrients than in the tropics. Trees shed leaves or needles, and the *humus layer* at the surface of the ground can accumulate faster than micro-organisms, worms, and insects can consume it. Rainwater or snowmelt that soaks through the rotting humus collects organic compounds that chelate and leach (dissolve) metallic cations from the underlying minerals and generally leave a silica-rich residue. Iron oxides and clay minerals that are washed from the surface layer accumulate a few feet down in the soil. Chemical weathering processes penetrate only a few feet into rock, but mechanical weathering,

FIGURE 2–5 *A South China landscape, idealized but still a reasonable interpretation of a limestone landscape under tropical humid weathering conditions. (Courtesy National Palace Museum, Taiwan, Republic of China.)*

FIGURE 2–6 *Aerial view of a central Pennsylvania countryside. Note the convex, rounded hilltops and the broad river valley. (Courtesy Pennsylvania Travel Development Bureau.)*

especially frost wedging, may extend considerably deeper. Precipitation and temperature being less intense than in the humid tropics, many dissolved compounds in the ground water can recombine as stable clay minerals rather than be swept away in solution. Clay minerals in the soil may prevent water from freely penetrating the ground. The resulting landscape develops broad, gentle, soil-covered slopes shaped by soil creep and stream erosion (Fig. 2–6). Ridge crests commonly have thin soils or exposed bedrock ledges.

Warm Arid Regions

Cold or hot, deserts are characterized by a deficiency of water. Chemical processes of all sorts are therefore inhibited. Weathering may cause granular rocks to crumble, but desert landscapes are characterized by jagged, angular landforms of mechanically fragmented or massive rocks, little smoothed by a cover of weathered debris. (Weathering and erosion in arid climates are described further in Chapter 4.)

For most of the time, ground water is drawn upward by capillary flow and surface evaporation. The ground water beneath an arid region has penetrated the ground elsewhere and moved laterally, often from cooler and more humid mountainsides. As the water rises in the ground and evaporates, salts accumulate. Under extreme aridity, salt encrusts the ground surface. Under less extreme conditions, only the least-soluble compounds such as calcium carbonate accumulate in the soil, and these in layers of concretions at a depth of several feet. The more soluble salts remain in solution.

In dry climates, polycrystalline rocks such as granite are less resistant to weathering attack than massive monomineralic rocks such as limestone and quartzite. Whereas in humid climates limestone is easily dissolved, and usually forms lowlands surrounded by highlands of other rocks, in a desert, limestone

FIGURE 2–7 *Arizona's Grand Canyon. An arid landscape, in which the strongest cliff-forming rock is the massive Redwall Limestone. (Courtesy Union Pacific Railroad.)*

becomes one of the most resistant rocks. Cliffs of Redwall Limestone, for instance, form the boldest escarpment in the Grand Canyon (Fig. 2–7).

Cold Regions

Cold regions are also among the driest parts of the Earth, for not only is precipitation slight, but the water that is present is normally in the solid form. Plants and soil micro-organisms are rare. A brief summer melt and intense cold make frost wedging the dominant process. The landscape, where not ice-covered, is a great sheet of rock rubble broken from exposed cliffs (Fig. 2–8).

FIGURE 2–8 *Wright Valley, South Victoria Land, Antarctica. The mean annual temperature is about −20°C. No soil covers the rock rubble on the taluses. (Courtesy C. Bull, Institute of Polar Studies, The Ohio State University.)*

Little chemical weathering can be detected, even by microscopic examination. Rock surfaces retain their true color, instead of being oxidized to brown or red. Thickly bedded limestone, as in arid regions, forms bold cliffs.

Soil

The term *soil* is used to describe the layer at the surface of the Earth that has been sufficiently weathered by physical, chemical, and biological processes so that it supports the growth of rooted plants. This is an agricultural definition, emphasizing that soil is a biologic as well as a geologic material. Engineers are less specific about their definition of a soil. To them, any loose, unconsolidated, or broken rock material at the surface of the Earth, whatever its origin, is soil. *Regolith* is a better term for the engineers' "soil." Geologists, like engineers, are inclined to regard soils only as weathered rock material. This is an overly narrow viewpoint, for soils are biologic entities as well, with immature and mature phases of life history.

The kind of soil that has formed at a place is a result of interactions between many materials and processes. The five principal factors of soil formation are: parent material (the local rock or transported rock debris), climate, vegetation, slope, and time. It is easy to visualize how the same weathering processes could react on different parent materials to give very different kinds of soil. Similarly, from the preceding description of climatic variants of the weathering processes, you can visualize how the same parent material might develop different soils under different climatic conditions and vegetational cover. The fourth factor, steepness of slopes, determines how quickly rainwater will drain from an area, or how deeply it will penetrate the ground, and thus this factor exerts a powerful control on the degree of weathering and the kind of soil. The fifth factor, time, is used in a relative rather than absolute sense. We speak of one soil as being older or more mature than another, not in a sense of years, but in the degree of development of diagnostic characteristics.

Soils are characterized by *horizons:* distinctive, successive layers, approximately parallel to the surface of the ground, that are produced by soil-forming processes. A *soil profile* is a vertical cross section of these horizons (Fig. 2–9). In the creation of a particular soil profile, oxidation may have discolored fragments of the parent rock to a depth of many feet, and carbonation may have leached all calcium carbonate from the upper several feet of the profile. Clay minerals and iron compounds leached from the upper foot of the profile may be concentrated in a layer at a depth of one to two feet. A dark layer of humus, or decaying vegetation, may form the uppermost six inches of the profile. By color, chemical tests, grain-size analyses and many other criteria, soil profiles can be subdivided into many horizons and subhorizons, and by a comparison of the nature and intensity of the horizons, soils can be classified.

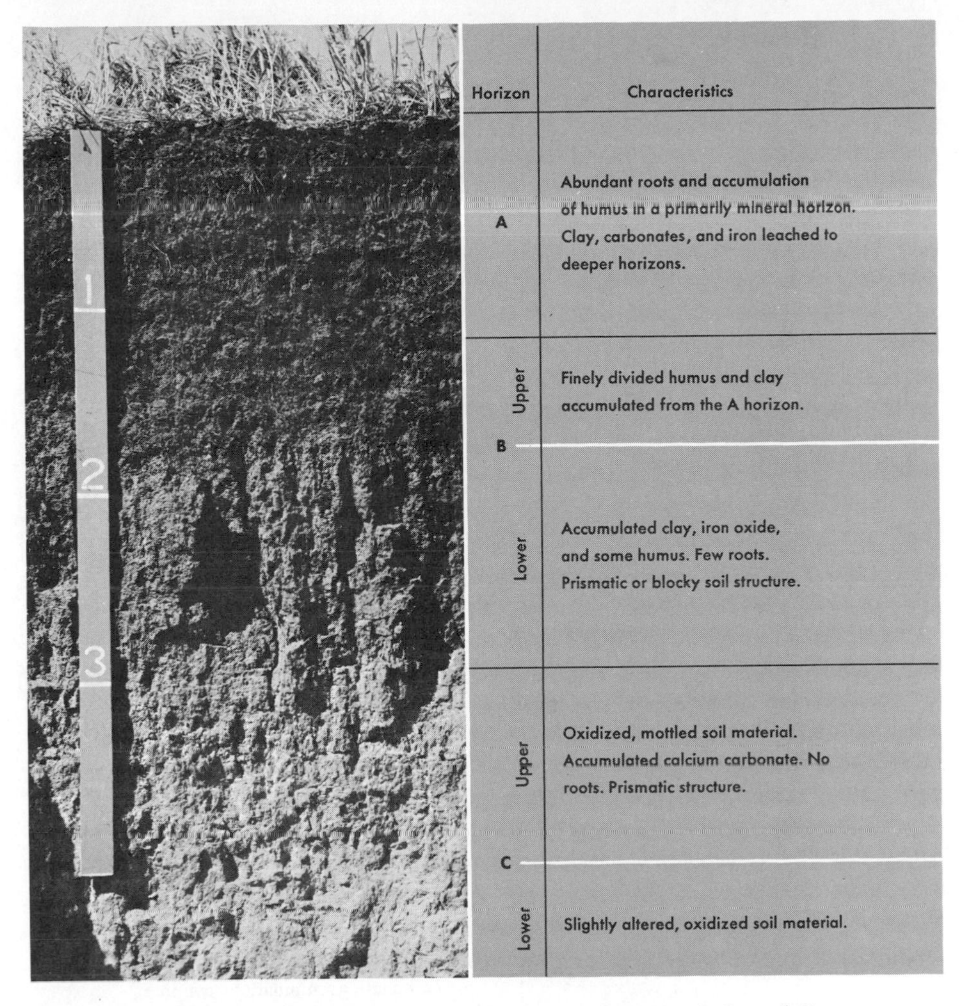

FIGURE 2–9 *A Mollisol soil profile, showing horizons A, B, and C.*

The five principal horizons, from the surface down into unaltered rock, are conventionally given the capital-letter symbols O, A, B, C, and R (Fig. 2–9). The O horizon is the organic upper horizon, dominated by fresh or partly decomposed organic material. The other four horizons are dominantly mineral in nature.

On the basis of the presence or degree of development of the various horizons in soil profiles, soils are grouped into ten orders. These orders are subdivided into an almost infinite variety of suborders, great groups, subgroups, families, and soil series. The classification that has been recently adopted by the United States Department of Agriculture, and will very likely be widely used in other countries because of its comprehensiveness, is presently called the "seventh approximation," for it represents the seventh full revision of a system of soil classification that has been under preparation by the Soil Survey Staff of the U.S. Department of Agriculture since 1951. The ten orders of the seventh approximation are listed in Table 2–1, in the confident belief that they represent the most comprehensive classification of soils and soil-forming processes yet devised.

No world map has yet been prepared of the distribution of the ten orders of the new soil classification, but when one is prepared, it will show that each soil order, except Entisols and Histosols, coincides with a distinctive climatic and vegetational region. The characteristics of the ten soil orders in Table 2–1 give a more detailed idea of regional and climatic variations in weathering processes than could be illustrated by the four examples of climatic control of weathering. The nearly 40 suborders of the new classification give even better definition to climatic controls.

Table 2–1

The Ten Soil Orders of the "Seventh Approximation" *

Name of Order	Derivation of Order Name	Character of the Soils
Entisol	Nonsense syllable "ent," from "recent"	Negligible differentiation of horizons in alluvium, frozen ground, desert sand, etc., in all climates.
Vertisol	L. *verto,* turn, invert	Clay-rich soils that hydrate and swell when wet, and crack on drying. Mostly in subhumid to arid regions.
Inceptisol	L. *inceptum,* beginning	Soils with only slight horizon development. Tundra soils, soils on new volcanic deposits, recently deglaciated areas, etc.
Aridisol	L. *aridus,* dry	Dry soils. Salt, gypsum, or carbonate accumulations common.
Mollisol	L. *mollis,* soft	Temperate grassland soils with a soft, organic-enriched, thick, dark surface layer.
Spodosol	Gr. *spodos,* wood ash	Humid forest soils. Mostly under conifers, with a diagnostic iron- or organic-enriched B horizon and commonly also an ashy-gray leached A horizon.
Alfisol	syllables from the chemical symbols Al, Fe	Clay-enriched B horizon, young soils commonly under deciduous forests.
Ultisol	L. *ultimus,* last	Humid temperate to tropical soils on old land surfaces, deeply weathered, red and yellow, clay-enriched soils.
Oxisol	F. *oxide,* oxide	Tropical and subtropical lateritic and bauxitic soils. Old, intensely weathered, nearly horizonless soils.
Histosol	Gr. *histos,* tissue	Bog soils, organic soils, peat, and muck. No climatic distinctions.

*From the Soil Survey Staff of the U.S. Department of Agriculture, 1960.

Soils tell a great deal about the history of weathering in a region. If climate has changed, or if forest has given way to grassland or farming, changes in the soil profile record the changing conditions. World climates have changed drastically and repeatedly in the last two million years or so, with glaciers expanding and contracting, and tropical regions becoming alternately wetter and drier. In many respects soils scientists are finding that soils, like the film in a faulty camera, may record multiple exposures of a whole succession of weathering processes. Sometimes it is difficult to distinguish in a soil profile the processes that are at work today and the processes that formerly weathered the landscape.

Conclusion

Weathering prepares the way for erosion. Long before stream valleys and gullies are eroded in a hillside, weathering has been localized along barely perceptible low places where water collects and penetrates the soil. B. T. Bunting in 1961 described an interesting example from northern England, where by mapping soils in detail, he showed that uphill from every little stream or ephemeral channel, the soil was unusually deep and well-developed along narrow, branching "seepage lines." As rainwater strikes the ground and begins its slow journey back to the sea, it collects in low places where its weathering activity is concentrated. Branching networks of more deeply weathered soils spread over all humid landscapes, preparing rock waste for subsequent removal by gravity and flowing water. This is the role of weathering in the geomorphic evolution of the land.

A. L. McAlester, in his book *The History of Life* in this series, discusses the theory that our oxygen-rich atmosphere is a byproduct of the evolution of photosynthetic life. It is an intriguing idea that weathering processes such as oxidation, which have been going on for enormously long geologic time, have been dependent on the presence of life on Earth. It is even possible to argue as follows: (1) The formation in large quantities of silica-rich igneous rocks such as granite requires weathering processes to concentrate sediments rich in quartz and clay minerals for melting; (2) the appropriate kinds of weathering can only operate in an oxygen-rich atmosphere; (3) atmospheric molecular oxygen is maintained by photosynthetic life processes; (4) therefore, photosynthetic life on Earth must be older than the oldest granite! These four logical steps cross the full spectrum of geologic research, and should be read as debatable propositions, rather than as assertions of truth.

3

Rock fragments in motion

When sufficient forces act on loose rock particles at the surface of the Earth, the particles move. This is true whether the particles are submicroscopic silica colloids in suspension in ground water or house-sized joint blocks that fall from cliffs. Ubiquitous gravity always adds a downward component to the motions produced by other forces. In general, then, when rock particles move, they move preferentially downhill.

The gravity component that acts parallel to a sloping surface is proportional to the sine of the slope angle (Fig. 3–1). The coefficient of sliding friction is equal to the ratio between the downslope component of gravity and the component of gravity acting perpendicular to the slope, or the tangent of the slope angle, when the particle is moving. Because few materials have a coefficient of friction greater than one, friction alone will not hold weathered rock on slopes greater than about 45° (Fig. 3–1). In fact, natural slopes steeper than about 40° are so rare that a British definition includes them in the category of cliffs. Debris-covered surfaces tend to have maximum slope angles of between 25° and 40°, depending on the shape and roughness of the particles.

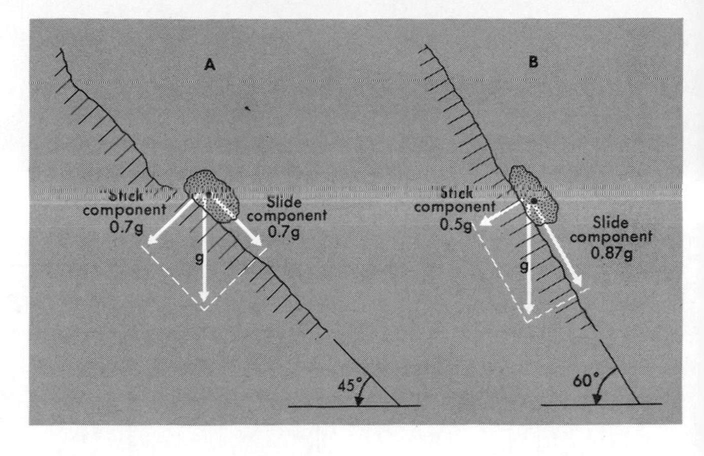

FIGURE 3–1 *Resolution of the gravitational force* g *that acts on a loose rock on a slope. On slope A, the component tending to slide the rock downhill and the component pressing against the hillside are both equal to about 0.7 g. On slope B the component tending to slide the rock downhill has increased to 0.87 g (sin 60°) and the component pressing against the hillside has decreased to 0.5 g. The rock is much more likely to slide on slope B than on slope A.*

Massive rock is strong enough to resist most surface forces that act on it. Mountains do not collapse and flow like taffy under their own weight. Only when rock has reacted with water and the atmosphere, or has been broken by mechanical stresses, can the fragments be mobilized. Weathering, then, is a necessary precondition for the movement of rock fragments down the surface of the land.

Mass-wasting

The collective term for all gravitational or downslope movements of weathered rock debris is *mass-wasting*. The term implies that gravity is the sole important force, and that no transporting medium such as wind, flowing water, ice, or molten lava is involved. Although flowing water is excluded from the process by definition, water nevertheless plays an important role in mass-wasting by reducing the coefficient of friction as a lubricant, and by increasing the weight of the weathered rock mass as a filler in pore spaces. Ice also may lubricate and increase the weight of rock debris and so hasten mass-wasting.

In 1938, C. F. S. Sharpe published a valuable little book on landslides and related phenomena. In it, he classified mass-wasting by an ingenious diagram from which Table 3–1 is simplified. The factors by which he subdivided mass-wasting were: first, the amount of included ice or water lubricant; second, the nature of the movement, whether sliding or falling as a coherent mass, or flowing by internal deformation; and third, the speed of the movement, ranging from the imperceptible to speeds of hundreds of feet per second under the full acceleration of gravity. The popular term *landslide* was not given any specific definition by Sharpe, but was retained as a useful general term for all rapid forms of mass-wasting. We will survey the various categories of mass-wasting, following the terminology of Table 3–1.

Table 3–1

Classification of Mass-wasting*

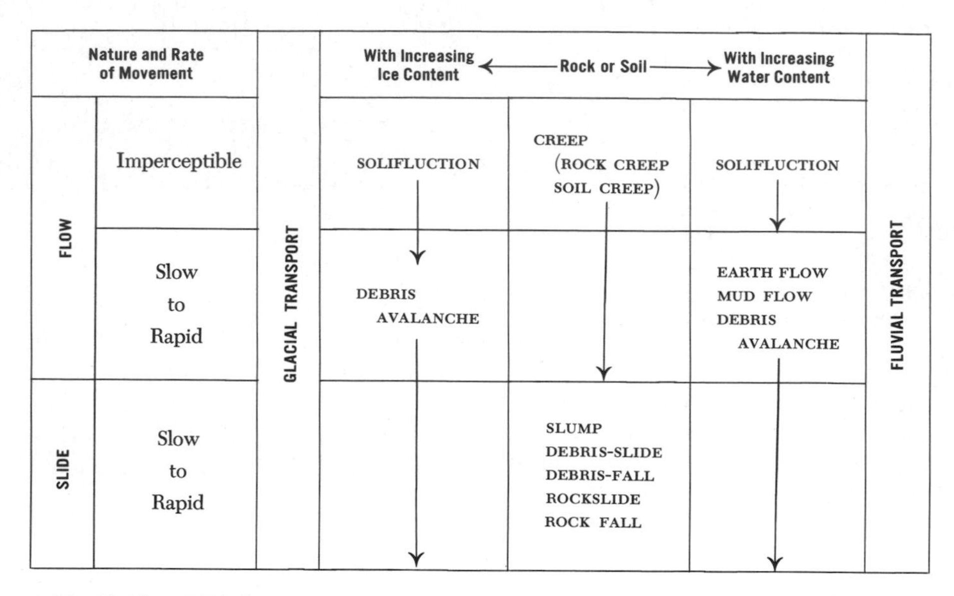

Nature and Rate of Movement		Glacial Transport	With Increasing Ice Content ←	Rock or Soil	→ With Increasing Water Content	Fluvial Transport
FLOW	Imperceptible	GLACIAL TRANSPORT	SOLIFLUCTION ↓	CREEP (ROCK CREEP SOIL CREEP) ↓	SOLIFLUCTION ↓	FLUVIAL TRANSPORT
FLOW	Slow to Rapid	GLACIAL TRANSPORT	DEBRIS AVALANCHE ↓	↓	EARTH FLOW MUD FLOW DEBRIS AVALANCHE ↓	FLUVIAL TRANSPORT
SLIDE	Slow to Rapid	GLACIAL TRANSPORT	↓	SLUMP DEBRIS-SLIDE DEBRIS-FALL ROCKSLIDE ROCK FALL	↓	FLUVIAL TRANSPORT

*Simpified from C.F.S. Sharpe, 1938.

Creep

The slowest mass-wasting is called *creep*. Depending on the material in motion, we speak of soil creep or rock creep. The rate of creep is greatest at the surface and gradually decreases to zero with depth. As a result, soil creep or rock creep does not shear across immobile rock at depth and is not capable of abrading a buried surface. Some of the phenomena of soil creep are sketched in Fig. 3–2. Grass may be able to maintain a continuous sod cover over an area of soil creep, for the downslope movement is measured in inches per year or less. Often, creep is detectable only from the uphill curvature of tree trunks that are many years old.

Soil creep is aided by expansion and contraction of soil by either freezing and thawing or wetting and drying. Volumetric expansion by any cause displaces particles toward the free face of the expanding mass, or perpendicular to the ground surface. On contraction, however, the particle is not pulled back into its former position, but settles with a gravitational component (Fig. 3–3). Only rarely are the cohesive forces of soil and water strong enough to draw particles back into the ground during contraction without some net downslope motion.

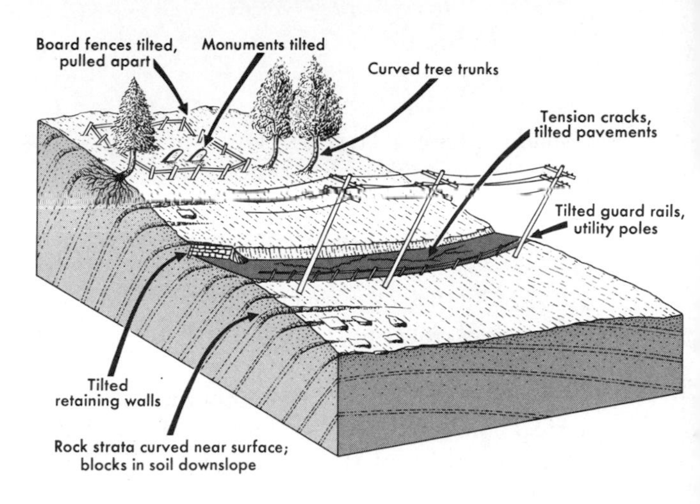

FIGURE 3–2 *Common effects of creep. Not all these features will be present in one place, but typically the detection of one will lead to the recognition of others.*

Solifluction, Earthflow, and Mudflow

If soil or regolith is saturated with water, the soggy mass may move downhill a few inches or a few feet per hour or per day. This type of movement is called *solifluction* (literally, "soil flow"). The process is especially common in subpolar regions, where the ground below a very shallow thaw-zone is permanently frozen. During the brief summer thaw, an "active layer" a few feet thick, composed of tundra peat, frost-shattered rock, and other weathered debris, may flow down slopes of almost negligible gradient, because meltwater saturates the active layer, but cannot penetrate the frozen ground beneath. The mass of solifluction debris may flow with a kind of rolling motion like the endless tread of a tracked vehicle. Arcuate ridges and troughs mark the toe, or lower part of the mass. Solifluction can be controlled by dewatering the moving mass through either natural or artificial means.

Solifluction is not a process restricted to frozen ground. It is a form of mass-wasting common wherever water cannot escape from a saturated surface layer of regolith. A clay hardpan in a soil or an impermeable bedrock layer can promote solifluction as effectively as a frozen substratum.

Earthflows or mudflows are features of mass-wasting very similar to solifluction. They are somewhat more rapid, and they commonly flow along valleys, whereas solifluction sheets or lobes cover an entire hillside with moving debris. Earthflows are common along the valley of the St. Lawrence River in

FIGURE 3–3 *Forces acting on a surface particle during expansion and contraction of soil on a slope.*

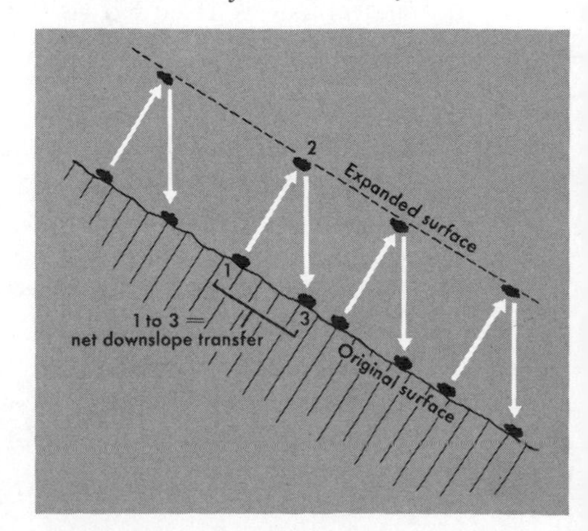

Canada, where a layer of clay is buried by sand. The clay, though water-saturated, is stable unless disturbed by explosive or earthquake shock or excessive artificial loading. Once the tenuous bonds between clay particles and water are broken, a mass liquefies spontaneously. The blasting of log jams in the river has been known to induce earthflows thousands of feet in width and up to 50 feet thick. Scars of ancient earthflows line the river terraces along the lower St. Lawrence River. The clay soils of Norway are notorious for similar earthflows.

During the March 27, 1964 earthquake at Anchorage, Alaska, the greatest devastation was caused not by the shaking, but by landslides that generally fit the earthflow category. Anchorage is built on a coastal plain of sand and gravel layers up to 60 feet in thickness that overlie a thick bed of clay. The vibration of the earthquake liquefied a sensitive layer in the clay, about 20 to 30 feet in thickness, which extended inland under the city at about sea level. Upper layers of stiffer clay and the overlying sand and gravel were literally floated seaward on the liquefied clay layer. Many of the slide areas moved nearly horizontally, with the liquefied clay extruding along the toe of the slide block as *pressure ridges*, and a downfaulted wedge, or *graben*, filling the gap at the back (Fig. 3–4).

The earthquake triggered the landslides at Anchorage, but the propelling force was gravity. One slide moved more than one-half mile across an intertidal mud flat into the ocean. Some slides continued to move for at least a minute after the earthquake shocks had ceased, demonstrating that the movement was not driven by the energy released in the earthquake. Many houses on the slide blocks, including a six-story apartment house, moved more than ten feet laterally with no structural damage, although all utility lines were severed. The soil at Anchorage in late March was still frozen to a depth of several feet; the frozen surface layer helped to hold the moving blocks in large pieces as they floated on the liquefied clay beneath, and prevented even greater damage and loss of life.

Earthflows and mudflows contain enough water to move in turbulent flow, and are known to erode channels as they flow. If any more water is involved, the movement is regarded as transportation by flowing water rather than mass-wasting. It must be emphasized that the various processes of mass-wasting

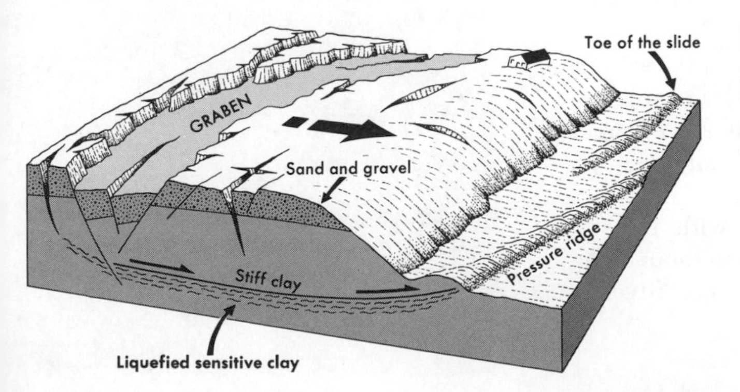

FIGURE 3–4 *Landslide on liquefied clay, typical of the damage at Anchorage, Alaska, during the earthquake of March 27, 1964. (After Hansen, 1965.)*

are transitional with one another and with the processes of water and glacier transportation. The terminology is not uniform, nor are terms rigorously defined.

Avalanche

The most rapid variety of flowing mass-movement is the *avalanche*, the very name of which is enough to frighten most mountain dwellers. In composition, an avalanche may range from entirely ice and snow to mostly rock debris. An avalanche usually begins with a free fall of a mass of rock or ice, which is pulverized on impact and flows at great speed, made fluid by the pressure-heated air and water entrapped in the mass.

One of the worst avalanches in history destroyed the region around Ranrahirca, Peru, on January 10, 1962, and by official estimate killed 3,500 people. Observers witnessed the entire catastrophe, from the time a huge ice cornice fell from an unnamed glacier near the peak of Huascaran (22,205 feet) until the debris came to rest against the opposite valley wall, nine miles away and two-and-one-half miles lower in altitude. The initial ice mass, of an estimated 3 million tons, tore loose other millions of tons of rock as it roared down the valley. The shock wave produced a noise like continuous growing thunder, and stripped hillsides bare of vegetation. Rocks and ice were pulverized by the turbulent flow. The avalanche required only seven minutes to travel 12 miles. It actually bounced from one side of a narrow gorge to the other at least five times before it emerged onto the fertile, heavily populated valley floor at the base of the mountain. As it spread one-and-one-half miles wide over the villages and fields of the valley, the mass slowed to an estimated 60 miles an hour and thinned to about 60 feet. When the avalanche stopped, air and water spouted from the settling debris. Later, melting blocks of ice created pockets of soft mud in the flow that were added hazards to the nearly hopeless search operations. There are rarely any survivors in the path of an avalanche.

Slump, Slide, and Fall

Under certain conditions a mass of rock or regolith may break loose from its bed and move downhill as a unit, sliding over the substratum along a definite surface. The least dramatic form of sliding mass-wasting is the *rotational slump*, whereby part of a hillside, usually of unconsolidated or weathered material, settles downward at its top and outward at its base (Fig. 3–5). The surface of failure beneath a slump block is spoon-shaped, concave upward or outward. The upper surface of a slump block commonly is tilted backwards, because the entire mass rotates as the lower part moves outward and downhill. Vegetation or even houses may be carried intact on the surface of a large slump block, as in the area near the graben in Fig. 3–4. At the toe of the slump an earth flow may emerge.

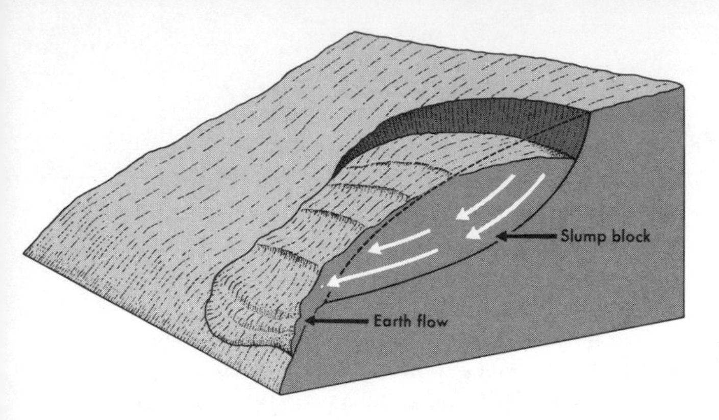

FIGURE 3–5 *Block diagram of a typical slump.*

Slumps may be caused by a stream or waves undercutting the foot of a slope. They are also a common result of faulty engineering design of cut embankments. They are sometimes controlled by loading the base of an unstable slope with a heavy layer of coarse rock rubble, which permits water to drain off the hill but offsets the weight of unstable earth higher on the slope.

Debris slides, rockslides, and *rock falls* are more dramatic forms of sliding mass-wasting. Large masses of unweathered rock may slide downhill along a sloping joint or bedding surface. Such a plane of weakness was probably involved in the Vaiont Reservoir disaster of October, 1963, in northern Italy. On the night of October 9, a rockslide 1.2 miles long, one mile wide, and over 500 feet thick moved suddenly down the south wall of the Vaiont Canyon, and completely filled the 875-foot-deep reservoir for 1.2 miles upstream from the dam to heights of 575 feet above the former water level. The movement took less than a minute, so rapid that the water in the reservoir was ejected 850 feet up the north canyon wall and propelled in great waves both upstream in the reservoir and downstream over the dam. The resulting floods killed 3,000 people, mostly around the town of Longarone, over one-and-one-half miles downstream from the dam and across a broad valley from the mouth of the Vaiont Canyon.

The geologic cross section of the reservoir is sketched in Fig. 3–6. The most obvious feature of the geology is the bowl-shaped structure of the rocks, which dip inward toward the valley axis from both sides. The rocks are mainly limestone, with thin clay layers at intervals. The limestone is full of caves and smaller solution channels, so that large amounts of rainwater can penetrate the rock and lubricate the clay layers. A whole series of natural and man-made factors contributed to the disaster. The principle ones were these. (1) The steeply dipping limestone beds and clay layers offered little frictional resistance to sliding. (2) All the rock types are inherently weak. (3) The river had eroded the steep inner canyon across the rock structure and removed lateral support long before the dam was built. (4) Two weeks of heavy rainfall had raised the water level in the cavernous rocks, and increased both fluid pressure and weight in the potential slide mass. (5) The high water level in the reservoir had saturated the lower part of the slide, decreased frictional resistance, and increased buoyancy.

The Vaiont Reservoir disaster was a surprise only in its severity. The canyon is marked by ancient landslides. In 1960 a smaller slide had occurred, and a pattern of cracks and slumps developed on the south valley wall that ultimately outlined the great slide of 1963. For six months prior to the slide, precise records of survey stations on the slide area showed that rock creep of one centimeter per week was in progress. By three weeks prior to the disaster, the rate of creep had increased to one centimeter per day; during the last week of heavy rains prior to the slide, the rate of creep had increased from 20 to 40 centimeters per day. Wild animals that had grazed on the south wall of the valley sensed the danger and moved away about October 1. On the night of the disaster, 20 technicians were on duty in the control building on the south abutment of the dam, and 40 more were in the hotel and office building on the north abutment, but no one survived who witnessed the actual slide. Desperate measures were underway to lower the reservoir level, but the rock creep was apparently pinching the reservoir and actually raising the water level in spite of open outlet gates. No earthquake or other "trigger" for the Vaiont slide has been identified. The rock along the principal slide plane simply failed under excessive weight and excessive water lubrication.

Mass-wasting and Landscapes

A rock cliff usually has at its base a sloping ramp of broken rock, or a *talus* (Fig. 2–8). The word refers to the slope, or the landform; the debris that forms a talus is called *sliderock*. As rocks fall from the cliff onto the top of the talus, they slide or roll until they lodge in place. Large blocks may roll to the foot of the talus, or they may be shattered and the fragments strewn down the slope. The entire sheet of sliderock gradually creeps downhill as material at the bottom is either weathered to more easily transported debris or eroded by flowing water.

A talus is a slope of transportation. That is, it assumes just the proper angle to maintain continuous downhill movement. If a talus is oversteepened by a chance fall of rock from the cliff, the downslope component of gravity is increased and creep and slide are accelerated until a stable slope angle is restored. If the base of a talus is undercut by a stream, sliderock moves down-

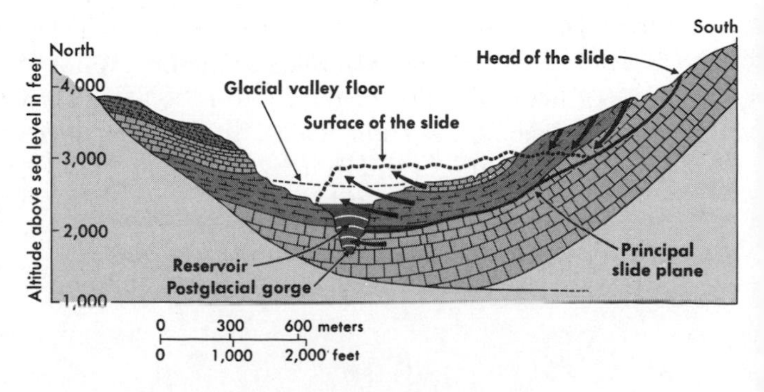

FIGURE 3–6 *Simplified geologic cross section of the Vaiont Reservoir, Italy. See Fig. 5–3 for a sketch of the inner gorge and dam. View is upstream (eastward). (Redrawn from Kiersch, 1964.)*

hill and exposes more of the cliff face at the top, from which more debris can fall to nourish the talus. When the cliff retreats until the talus extends entirely to the top of the hill, the supply of sliderock ceases, and the talus becomes covered with soil and vegetation and loses its distinct identity.

It is unfortunate for our perspective that an analysis of mass-wasting becomes a description of major human disasters. Since avalanches and rockslides kill thousands of people, they deserve intensive study to save lives through better prediction, control, and warning procedures. But these spectacles are restricted to areas of high relief, steep slopes, and local special conditions. In much less conspicuous fashion, weathered rock and soil slowly creep down all slopes, whether under grass, forest, desert sagebrush, or tundra. Creep is surely the dominant process of mass-wasting, but the evidence for it is subtle and often overlooked. Every tilted sidewalk slab, cracked pavement, and slumped embankment announces that creep is in progress.

Slope Development and Maintenance

Most of a landscape consists of curved, sloping surfaces, largely shaped by mass-wasting. How these slopes form, how they are maintained, and how they change with time are major topics of geomorphic research. Slopes are exceptionally difficult to study, for both in processes and in form they are transitional. Recall the arbitrary definitions of mass-wasting: Add a little water, and soil creep becomes earth flow; add a little more water, and earthflow becomes a muddy stream.

It is surprisingly difficult even to describe a natural slope geometrically. We are commonly provided with only a profile surveyed down the steepest part of a hillside as a description of the slope, but obviously a profile measured along the crest of a descending ridge is going to have a significance very different from that of a similarly shaped profile measured along the bed of an adjacent gulley. Slopes are irregular surfaces that cannot be described by simple mathematical equations. The best topographic maps are only approximations of the infinite irregularities of hillsides. We do not yet know what degree of irregularity is significant in the stability of slopes, so we are never sure that we are measuring the correct angles and distances.

Why are landscapes so complexly curved? Two broad classes of formative processes can be distinguished as the beginning of an answer to that question. First, internal forces of the Earth raise areas above sea level as *tectonic* landforms such as mountain chains. The uplift is never regular. Rocks are folded or faulted into ridges and troughs that alone would give great diversity to surface slopes. Second, we rarely see true tectonic landforms. Most landscapes are *erosional.* As soon as land rises above the sea, gravity and flowing water begin to tear it down again. The geomorphology machine grinds into action. Flowing water cuts a network of valleys over the land, and every valley that is eroded produces two new valley-side slopes. The interaction of erosional processes (including for the moment weathering and mass-wasting) with tectonic deformation produces the infinite variety of landscape slopes. It is a useful mental exercise to realize that under the sea or on the Moon, erosional processes and even tectonic processes are not the same as on subaerial landscapes. As we begin to explore the sea floor and the Moon, we must keep open minds about the origins of the forms we see.

There have been two philosophies about studying slopes. An older school, exemplified by W. M. Davis, deduced the systematic changes of slope form that would accompany long-continued subaerial weathering and erosion. Because landscape evolution is too slow to be witnessed, deductions concerning the changes of slope form with time were based on assumptions that could not be tested until we had isotopic techniques for dating old land surfaces. No wonder that the deductive approach to slope analysis has enriched geologic literature with some remarkably opinionated and authoritarian writings. A majority of deductive geomorphologists have held that slopes, especially in humid regions, become lower and more broadly rounded with time. A vocal minority have insisted that slopes are stable forms with angles controlled by rock type and weathering process, and once a stable slope is evolved, it persists through time, migrating backward parallel to itself unless it is eliminated by the intersection of other slopes. We will consider these matters further in Chapter 5.

Another group of geomorphologists have concerned themselves with the empirical description of slopes. With less regard for theoretical projection into the future, they have studied the processes of slope formation and the geometry of slopes. Innumerable slope profiles and descriptive texts have been published, but empirical study has suffered from the lack of a guiding theory. Only dedicated and persistent workers continue to scramble up and down hills with measuring tapes and levels.

Some progress has been made, however, by both deductive and empirical methods of study. We now recognize that slope profiles generally have an upper segment convex to the sky, and a lower concave segment (Fig. 3–7A), that some slope profiles show a straight segment between the upper and lower curves (Fig. 3–7B), and that if a cliff interrupts the slope, an additional segment

FIGURE 3–7 *Profiles of slopes.*

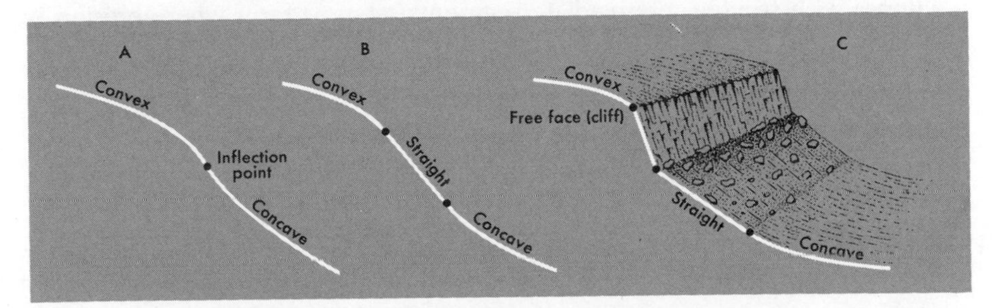

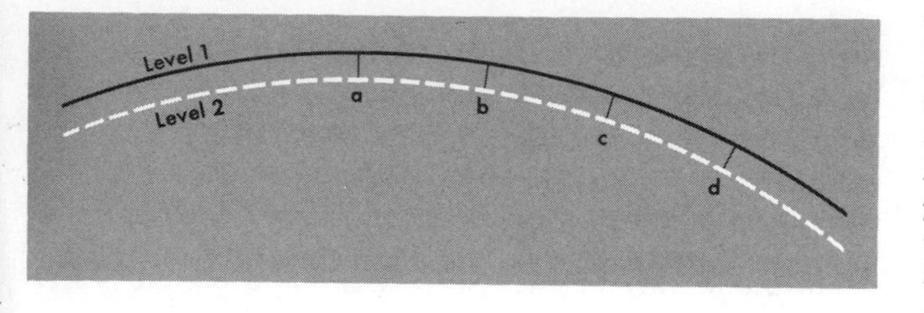

FIGURE 3–8 *G. K. Gilbert's explanation of the summit convexity. In order that the surface be lowered from Level 1 to Level 2 uniformly, progressively greater masses must move downslope past b, c, and d, in a given time. Hence, slopes must steepen radially from the summit.*

marked by free fall of weathered debris is introduced into the profile above the straight segment (Fig. 3–7C). The straight segment of a slope below a free face is usually a talus.

In general, the upper convex curvature of the profile is controlled by mass-wasting, especially creep. G. K. Gilbert in 1909 offered an explanation of summit convexity that is largely deductive but is still the best available explanation. He assumed a uniform thickness of soil or regolith over the convex surface (Fig. 3–8). If, in a given interval of time, a uniform thickness of the weathered material is removed from the entire summit area, progessively larger quantities must move through cross sections progressively farther downhill. In other words, with the stated assumptions, the amount of material that creeps past any point is proportional to the distance of the point from the summit. As creep is primarily a gravitational phenomenon, the slope angle must increase radially from the summit in order to move the progressively greater amount of debris. The summit curvature becomes convex to the sky. Measurements of soil creep verify that it is the dominant process on vegetated upper hillsides.

On lower slopes, transportation by flowing water assumes dominance over creep. Two little rills flowing down a bare hillside during a rainstorm require a certain slope to keep flowing with their suspended-sediment load. When the two rills join, the resulting rivulet has a greater proportional increase of mass than its increase in wetted surface area. Friction is reduced in proportion to the discharge, and the larger trickle of water can transport the joint loads of the two lesser streams with no loss of velocity, but on a more gentle slope. Thus, slopes controlled by *rainwash*, *sheetwash*, or *rillwash* are generally concave skyward. At some position on a slope, rainwash becomes dominant over soil creep and the slope profile inflects from convex near the top to concave near the base.

Slopes with straight intermediate segments seem to form where erosion is unusually rapid. In the extreme examples of gullies cut on artificial embankments by single intense storms, most of the slopes are straight. Natural landscapes scored by closely spaced, V-shaped gullies with straight sides that intersect as knife-edged ridges are called *badlands*.

Thus far we have considered only slope profiles. Even from profiles alone, we can see how soil creep near the top of a slope will increase the gradient downhill until rainwater begins to flow over the surface instead of penetrating and lubricating the creeping soil. At that level on the hillside, sheetwash and slope concavity begin.

Slopes also curve in directions other than downhill, and these other curvatures also affect water movement. Where contour lines bulge convexly outward on a hillside around sloping spurs or *noses,* water is spread laterally as it flows downhill. Noses and ridge crests tend to be drier than adjacent *hollows,* where the contour lines swing concavely into the hill. Concave contours tend to gather water from a large area higher on the slope. The heads of streams are localized downhill from hollows.

We can combine profile curvature and contour curvature into a single diagrammatic classification of slopes (Fig. 3–9). The horizontal axis of the diagram divides "water-gathering" slopes with concave contours (quadrants I and II) from "water-spreading" slopes with convex contours (quadrants III and IV). The vertical axis of the diagram separates slopes with convex profiles dominated by creep (quadrants II and III) from those with concave profiles dominated by rainwash (quadrants I and IV). F. R. Troeh, who published this ingenious classification of slopes in 1965, found that he could place almost any land surface into one of the four quadrants of the diagram. The only exceptions are saddle-shaped surfaces, which require a higher order of mathematical analysis. Horizontal surfaces or straight, flat slopes plot on the axes of the

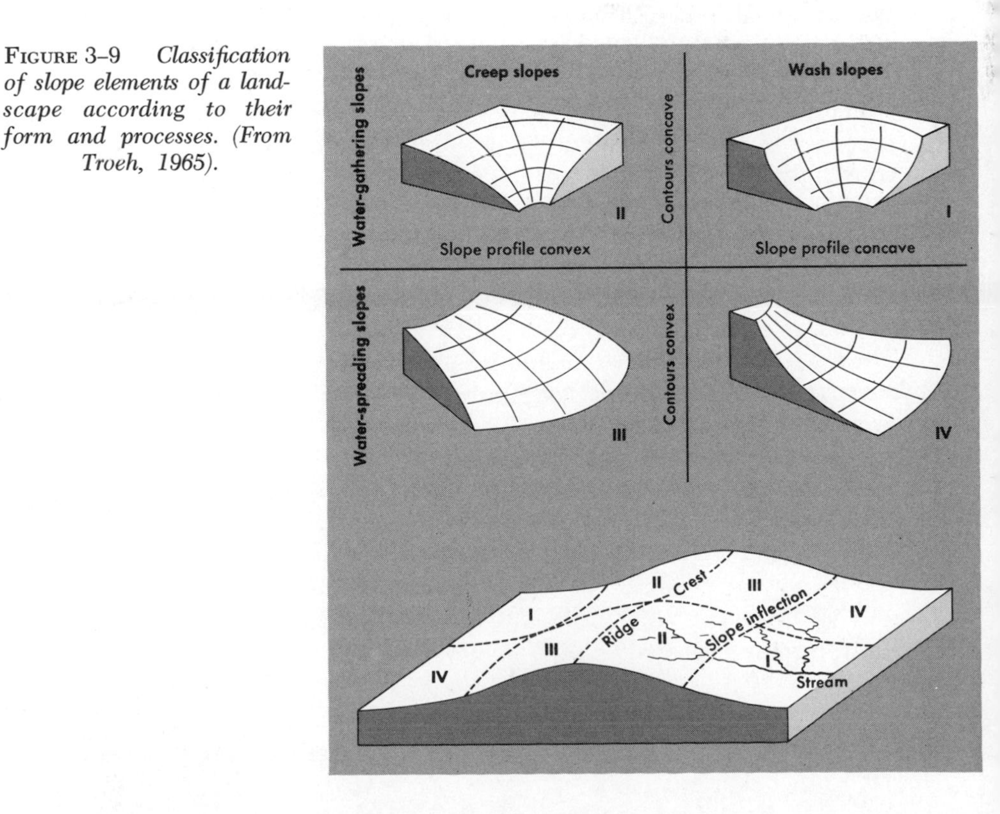

FIGURE 3–9 *Classification of slope elements of a landscape according to their form and processes. (From Troeh, 1965).*

diagram. At the bottom of Fig. 3–9 is a block diagram of a sinuous ridge, labeled to show how a hillside can be subdivided into the component slope elements.

Each of the slope elements of Fig. 3–9 can be represented mathematically by a quadratic equation. Each surface is generated by rotating a segment of a parabola around a vertical axis. In perfecting his classification, Troeh surveyed agricultural lands around Cornell University, and using an electronic computer, he calculated the best-fitting paraboloid of rotation for each small area of surveyed land. He discovered that in areas where the actual land surface could be represented by a single quadratic equation with vertical deviations no greater than four to six inches, a single soil type was represented. Where the land surface departed from the calculated surface by more than six inches, a new soil type appeared. Usually the difference in soil was controlled by slightly better or worse drainage. Troeh's conclusions strongly support the first theme listed in the introduction of this book: Continuity of landform indicates continuity of process.

Ordinarily, a landscape is composed of small slope elements, each reacting in a particular way to the local effectiveness of weathering, mass-wasting, and erosion. All elements are related, however, for an accidental disequilibrium in any part of a slope affects adjacent segments above and below the site of the accident. An animal burrowing on a hillside raises a mound of loose soil on the downhill slope. The slope is locally oversteepened, and the debris is rapidly spread downhill. Simultaneously, rainwash from uphill is trapped by the burrow, and creep is accelerated by the undermining. A dynamically stable, or *graded,* slope is an example of an *open physical system* through which both energy and matter move, a system that tends by self-regulating processes to maintain itself in the most efficient possible configuration. Slopes constantly change, but always tend toward some central, graded state appropriate to the environment of the moment.

4

Streams and channels

Water, flowing down to the sea over the face of the land, is the dominant agent of landscape alteration. Surface weathering and ground-water solution provide a load for flowing streams, and mass-wasting may dump great quantities of rock debris at the foot of a slope, but eventually rivers must carry all but a small fraction of the total rock waste from the lands to the sea. Wind, glaciers, ocean waves, and all other erosional agencies do only minor amounts of work as compared to rivers. Therefore, to understand how landscapes change, we must first understand how rivers do their work.

During a rain, the water gathers into rills down hillsides, and cuts a myriad of ephemeral channels ("rillwash"). At the end of the rain, the water soaks into the channel floors and locally intensifies weathering. The little channels quickly fade through creep, but beneath each one a little more rock has been weathered and prepared for removal. Below water-gathering slopes, runoff continues for sometime after a rain. If the *catchment area*, or *watershed*, of a slope is large enough, a permanent stream may form near its base, fed from both surface and subsurface flow.

The proportion of surface and subsurface water that feeds a stream varies greatly with climate, soil type, bedrock, slope, vegetation, and many other factors. One estimate is that one-eighth of the annual runoff of the hydrologic cycle (Fig. 1–3) goes directly overland to the sea, while seven-eighths of the water goes underground at least briefly. If you recall how rapidly infiltrating water reacts with minerals (Chapter 2), you will understand why "pure" spring water closely reflects the chemistry of the local rocks. Even as river water begins its downhill flow, its chemical energy has been largely expended.

Dynamics of Flowing Water

The patterns of stream channels are varied and complex. Like the branches of a tree, a drainage system repeatedly subdivides upstream into ever smaller channels. The finger-tip tributaries of a river system, those that emerge from the ground and flow to a junction without receiving any tributaries themselves, are called *first-order streams.* Of course, if you were to study a slope during a rain, you would find that each first-order stream is actually the trunk river of a complex but almost microscopic drainage network that functions only while the rain is falling.

Drainage networks have highly probable patterns when analyzed statistically. A good model of a river system can be generated by a "random walk" game. Start a number of markers at equal spaces along one edge of a piece of graph paper. Move the markers either ahead, left, or right, one space at a time, by the cast of a die. One restraint to random movement is that a marker cannot move backward. This is the game equivalent of gravity. If a marker intersects the path of another marker, it must follow the previously defined path from that point. This rule of the game represents the cohesiveness of water. The paths of the markers will trace drainage nets, in which first-order tributaries join to form second-order streams, and so on until either all paths have merged into a single master stream, or diverged beyond any probable junction. The frequency with which both game markers and streams join, the average length of path between successive junctions, and other parameters of the networks depend only on the original rules. If one part of the net is known, other parts can be predicted. In nature, the "rules" include the structure of the bedrock, the tectonic slopes, and the previous history of erosion. Nature's rules are both complex and subtle.

One of the interesting aspects of stream flow is that as the quantity of water in a stream increases, the down-valley slope of the water surface decreases. As an empirical rule, *slope is an inverse function of discharge.* No adequate theory is available to explain this rule, but its validity is based on direct observation. Apparently, water flows more efficiently in larger channels, and therefore requires less slope to maintain its velocity.

The inverse relationship between slope and discharge can be illustrated by a practical application, if not explained by a general theory. In irrigation systems, each ditch must slope steeply enough to keep water moving and keep mud from settling in the channel, but not too steeply to cause the water to erode the banks of the ditch. A delicate energy balance must be maintained. Through centuries of trial-and-error and experiment, men have learned that successively smaller distributary irrigation ditches must be given successively steeper gradients to keep the water moving at the proper speed. In an irrigation system the discharge decreases downstream, as the water from a single large feeder canal is divided and subdivided among districts, farms, and fields. The inverse rule between slope and discharge holds, even in this reversal of the pattern of a normal river system.

In humid regions, the discharge of rivers increases downstream. Not all of the annual runoff from the land is poured into rivers at the heads of first-order tributaries, equally spaced from river mouths like the markers at the start of a game. Water can be added anywhere along a river from surface runoff and from subsurface seepage. Rivers in humid regions are called *effluent* because they receive contributions of ground water. Rivers in arid regions generally lose water to the ground in addition to losing it by evaporation, and often they dry up entirely without reaching the sea. These are called *influent* streams; their distinctive channel characteristics will be described later in this chapter.

Where rivers enter the sea, the potential energy of the falling water reaches zero. No further conversion of potential energy to stream work is possible, so sea level, and its projection under the land, is called the *ultimate base level* of stream erosion. Actually, most streams enter the sea with a considerable velocity, and therefore have kinetic energy available to erode their channels well below sea level, but this observation does not invalidate the use of sea level as a reference level for the limit of potential energy conversion. We will see later that there are also *local or temporary base levels* that delay stream erosion, but never halt it.

Because rivers have their greatest discharge and therefore their most gentle gradients near their mouths, they enter the sea at smooth, almost undetectable tangents. The Mississippi River at New Orleans is at sea level 107 miles upstream from its mouth (actually mouths) in the Gulf of Mexico. The mighty Amazon River is only about 20 feet above tidewater at Obidos, Brazil, 500 miles from its mouths.

Hydraulic Geometry of Stream Channels

For many years, agencies of governments have maintained *gaging stations* along rivers all over the world. At these stations, the water-surface level, channel shape, stream velocity, amount of dissolved and suspended mineral matter, and other variables are periodically or continuously recorded. The discharge of a stream is also measured, not directly, but by multiplying the cross-section area of the channel at the gaging station by the average velocity of the current. Discharge is expressed in cubic feet per second (cross-section area in square feet $\times$ velocity in feet per second) or equivalent units.

The records of stream gaging stations are essential for the prediction of potential flood damage, stream pollution, and other disasters. These records also provide a voluminous history of stream flow. In 1953, a team of geologists and engineers from the U.S. Geological Survey published an analysis of thousands of measurements from stream gaging stations all over the United States. They called their analysis of the relationships between stream discharge, channel shape, sediment load, and slope the *hydraulic geometry of stream channels*. The work is continuing, and is one of the most fruitful and potentially useful areas of geologic research.

The training of the two innovators of the term "hydraulic geometry" reveals the complexity of the research. L. B. Leopold has degrees in civil engineering, meteorology, and geology. Thomas Maddock has a degree in civil engineering and specializes in hydraulic engineering. They needed all this training and experience to recognize the geomorphic significance of variations in stream flow and channel shape.

The first step in analyzing the hydraulic geometry of stream channels was to study the changes in channel width and depth, stream velocity, and suspended load at selected gaging stations during conditions ranging from low flow to bank-full discharge and flood. Over a wide range of conditions, it was found that width, depth, velocity, and suspended load increase as simple power functions of discharge. That is, all increase as some small, positive, exponential function of discharge. Some of the gratifyingly simple equations are:

$$w = aQ^b \qquad d = cQ^f \qquad v = kQ^m$$

where Q = water discharge, w = width, d = mean depth, and v = mean velocity. The changes of load with discharge will be described later in this chapter.

The numerical values of the arithmetic constants a, c, and k are not very significant for the hydraulic geometry of streams, but the numerical values of the exponents b, f, and m are very important. Leopold and Maddock found that the average of 20 representative gaging stations in central and southwestern United States gave values of the exponents b, f, and m, as follows:

$$b = 0.26 \qquad f = 0.40 \qquad m = 0.34$$

These values signify that as the discharge of water past a gaging station increases, perhaps during a flood, the width of the channel increases approximately as the fourth root of discharge ($w = aQ^{0.26}$), the mean depth increases nearly as the square root of discharge ($d = cQ^{0.40}$), and the velocity increases about as the cube root of discharge ($v = kQ^{0.34}$). Channel width, depth, and current velocity all increase at gaging stations during rising water. This conclusion is no surprise to anyone who has seen a river in flood, but the regularity of the changes is significant.

More surprising are the results of comparing the changes in channel shape and stream velocity in a downstream direction. We have seen that river discharge in humid areas increases downstream. When the mean annual discharge of many rivers past successive gaging stations was compared to the width, depth, and velocity at each station, the same equations were found to apply that had been derived for changes in flow past a single point. *As discharge of a river increases downstream, channel width, channel depth, and current velocity all increase.*

Everyone knows that rivers get both wider and deeper as they grow larger downstream, but until Leopold and Maddock published their work, no one had guessed that average current velocity also increases downstream. The conclusion violates our poetic impressions about wild, rapidly flowing mountain streams and deep, wide, placid rivers like the Mississippi. We do not immediately realize that much of the current in a mountain torrent flows in circular eddies, with almost as much backward as forward motion.

Figure 4–1 reproduces an example of the evidence that mean current velocity increases downstream with discharge, first published by L. B. Leopold in 1953. Both velocity and discharge are plotted on logarithmic scales to show as a straight line the exponential relationship between the variables.

The numerical values of the two exponents b and m are not the same for

FIGURE 4–1 *Velocity and discharge of the Yellowstone-Missouri-Mississippi River system, demonstrating that average velocity in a river increases downstream with increasing discharge. Additional gaging stations, unnamed, are shown by X. (Data from Leopold and Maddock, 1953.)*

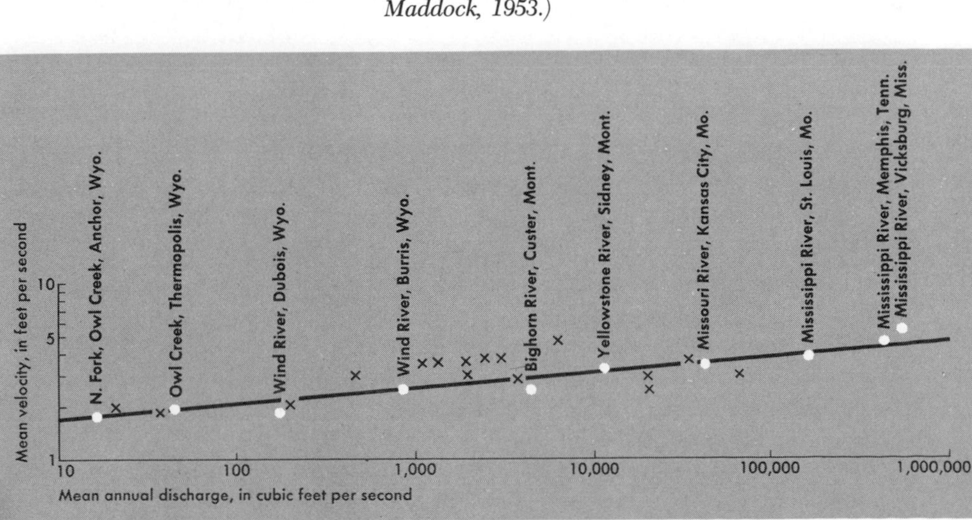

changes downstream and for changes with discharge past a point. In the downstream direction, the average values for the exponents were found to be:

$$b = 0.5 \qquad f = 0.4 \qquad m = 0.1$$

In the downstream direction, channel width increases most rapidly with discharge, depth next most rapidly, and mean velocity increases only slightly, although it definitely increases. It is believed that the increasing depth downstream permits more efficient flow in a river, and overcompensates for the decreasing slope, thus providing a slight net increase in velocity at mean annual discharge.

A mathematical test of the hydraulic geometry equations suggests useful applications of the principles. We defined discharge as area times velocity, or $Q = wdv$. If:

$$w = aQ^b \qquad d = cQ^f \qquad v = kQ^m$$

then by substitution:

$$Q = (aQ^b)(cQ^f)(kQ^m)$$

or:

$$Q = ackQ^{b+f+m}$$

it follows that:

$$a \times c \times k = 1.0$$

and:

$$b + f + m = 1.0$$

We need not be concerned with the arithmetic constants a, c, and k, but it is interesting to verify that in the examples given for both single gaging stations and downstream, $b + f + m = 1.0$.

We may be approaching the day when we will be able to predict, by hydraulic geometry, the flow characteristics of rivers from a bare minimum of observational data. Great savings in equipment and time will be possible if a small number of representative gaging stations, perhaps with automatic recording and transmitting equipment, will provide all the information we need for reservoir recharge, flood prediction, irrigation, pollution abatement, river navigation, and many other human uses of rivers. The U.S. Geological Survey now maintains about 4,300 gaging stations as the *hydrologic network*, or *surface-water basic-data network*, of the United States. Additional thousands of gaging stations are also maintained for specific needs of regional water management.

Transportation and Erosion by Streams

Streams transport almost all of the weathered rock debris from the land to the sea. Winds and glaciers transport only a small amount by comparison, and the waves that erode the edges of the land are effective only in a narrow fringing shore zone. Streams carry their loads of weathered sediment in three ways. Some particles (silt and clay, collectively called *mud* when wet and *dust* when dry) are small enough to be kept in suspension by turbulent water flow. These particles are the *suspended-sediment load.* Larger (sand and gravel sizes) or heavy rock fragments roll, slide, or bounce along the bed of a stream. These form the *bed load* of the stream. The weathered constituents of rocks that are carried in chemical solution make up the *dissolved load.*

Of the various components of stream load, only the suspended-sediment load is at all well known. At certain gaging stations, standard volumes of water are routinely collected, and the suspended sediment is separated, dried, and weighed. The weight of sediment per unit volume of water is multiplied by the discharge at the time of sampling to give the sediment load in tons per day or equivalent units.

No satisfactory method for measuring bed load has been devised. Any sampling device placed on the bed of a stream deflects the current and changes the transportation properties along the bed. The bed load of streams is commonly assumed to be about 10 per cent of the suspended load, although in some rivers it is over half of the total load.

The dissolved load is calculated from chemical analyses of the water and the discharge values. The special chemical analyses are too expensive and time-consuming to be done routinely except at a few stations, so we estimate the regional patterns of dissolved load by studying a very few, hopefully typical, rivers.

The hydraulic geometry of streams involves suspended-sediment load and bed load as well as discharge, width, depth, velocity, and slope. As far as we know, the dissolved load does not affect the physical properties of flowing water. The equation relating suspended-sediment load to discharge is similar in form to the equations for width, depth, and velocity. The equation given by Leopold and Maddock is $L = pQ^j$, where L is suspended-sediment load, Q is discharge, and p and j are numerical constants.

In general, as the discharge increases at a gaging station, the suspended-sediment load increases. Values for the exponent j range from 2.0 to 3.0. These large exponential values mean that as discharge at a station increases tenfold, the suspended load may increase a hundredfold to a thousandfold! Figure 4–2 is a typical graph of suspended-sediment load compared with discharge. The suspended-sediment load at a station increases much more rapidly with discharge than either channel width or depth, therefore the enlargement of the channel by erosion cannot account for all of the increased load. Most of the suspended sediment comes from the watershed upstream from the gaging station. The sediment is newly delivered to the stream by mass-wasting and rill-wash during the same rains or snow-melts that swell the discharge of the river.

The change of suspended-sediment load downstream with increasing discharge has not been measured directly, but it can be estimated indirectly from other parameters of hydraulic geometry. The value for exponent j in a downstream direction is 0.8, which implies that the total suspended-sediment load increases downstream slightly less rapidly than discharge, and the concentration of suspended sediment becomes more dilute toward the river mouth.

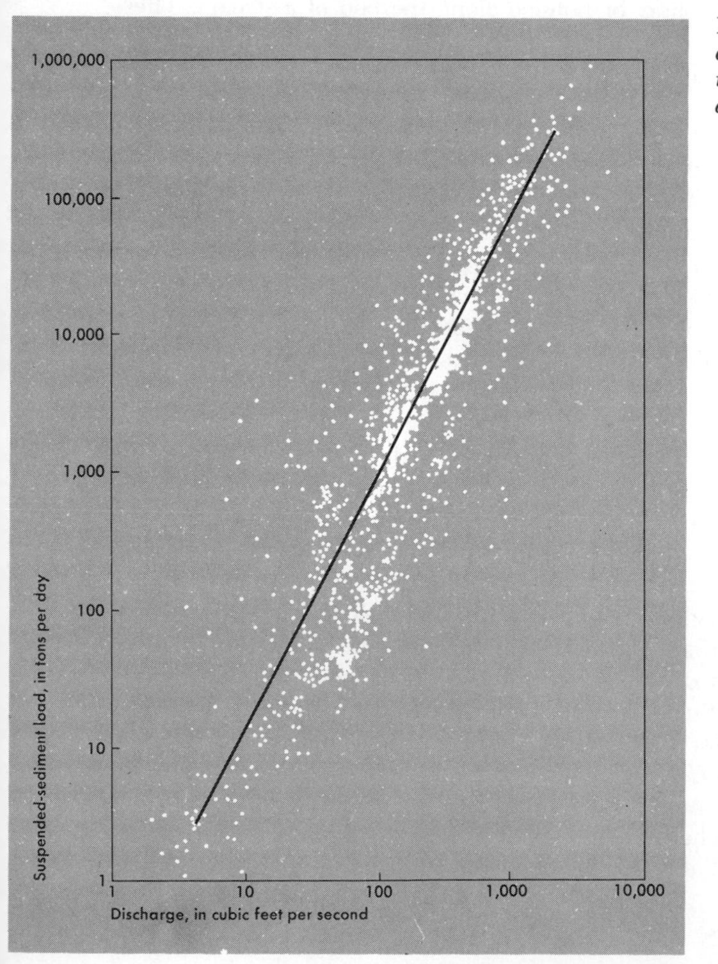

FIGURE 4–2 *Relationship of suspended-sediment load to discharge, Powder River at Arvada, Wyoming. (Leopold and Maddock, 1953.)*

Measurements of channel shape and suspended-sediment load confirm that streams move most of their loads during times of high discharge. As the water rises during a flood, the channel is scoured deeper and wider and additional quantities of suspended sediment are added to the river from the watershed slopes. Velocity increases, enabling the stream to transport larger-sized fragments and a greater total load than it can move at lower discharges. As the flood subsides and velocity slackens, the load is dropped again, rebuilding the channel bottom to its former configuration. We might believe that no permanent change had taken place in the river channel during a period of high discharge, unless we realize that the mud, sand, and gravel that previously formed the stream bed at a point has been moved downstream toward the sea, and has been replaced by new sediment from upstream. The geomorphology machine has surged ahead slightly.

During low water, or periods of low discharge, much of the nondissolved sediment load of rivers must be dropped. The sediment is called *alluvium*. Well-established rivers usually have their valley floors covered with alluvium, into which the low-water channel is carved. The surface of the alluvium, from the banks of the low-water channel to the base of the valley walls, is called the *flood plain* of a river. The name is appropriate, for in flood the entire flood plain becomes the bed of the river. Many rivers regularly flood over the banks of the low-water channel at an interval of from one to three years. The "mean annual flood" is a normal feature of rivers that are not artificially confined to their low-water channels.

The flood plain and the alluvium that compose it are vital to a river in several ways. When a river goes over its banks in flood, the channel width is suddenly increased to the full width of the flood plain. As discharge is the product of width, depth, and velocity ($Q = wdv$), a great increase in width can accommodate the increased discharge of a flood with only slight increases in velocity and channel depth. The velocity of the water over the flood plain outside the main channel may even be low enough so that sediment can settle out of suspension and add to the alluvium. People who live on flood plains know that a single flood may deposit a foot or more of muddy alluvium on their fields and in their houses.

A river with a well-developed flood plain flows in broad, regular curves called *meanders*. It is another of the little-understood basic properties of water to flow in curving arcs. Sediment-free streams flowing over glaciers during the summer develop meandering channels on pure ice. Even the Gulf Stream is believed to meander like a great river in the Atlantic Ocean. Streams that flow in erodible materials tend to form meanders, and flood-plain alluvium is ideally erodible. The same stream that transported the alluvium so far toward the sea obviously is capable of eroding it again and transporting it still farther.

Rivers undercut their banks along the outside curve of meanders and build sand or gravel shoals called *point bars* along the inside of the bends (Fig. 4–3). The eroded bank material is usually swept downstream a short distance to the

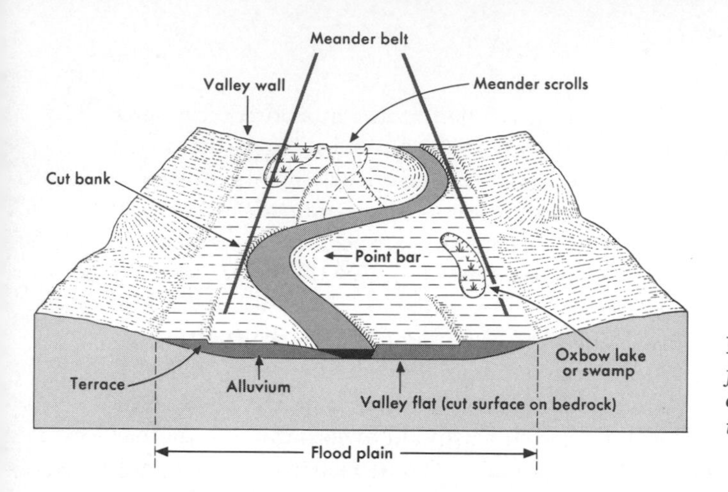

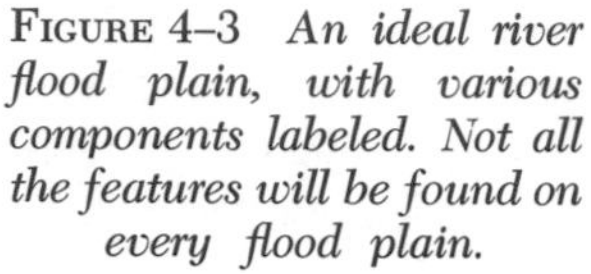
FIGURE 4–3 *An ideal river flood plain, with various components labeled. Not all the features will be found on every flood plain.*

next point bar. In time, a meandering river will swing laterally across the entire flood plain, and at the same time the meanders will migrate downstream. Flood plains are marked by arc-shaped depressions of former meanders, or *meander scrolls.* Cutoff meander segments form *oxbow lakes* on the flood plain. Small *terraces* cut in the alluvium mark successive levels of flooding.

From time to time, each particle of alluvium in the flood plain is re-exposed by the meandering channel. Small particles that were originally dropped at a low velocity are swept away in suspension. Rock fragments in the alluvium that have weathered smaller since their initial deposition are removed. The alluvial fragments are continuously size-sorted, rounded, and reduced in size by abrasion, as they are moved downhill to the sea. One estimate is that the average storage time for alluvium on a flood plain is on the order of 1,000 years.

When large, muddy rivers flood, much of the excess suspended load settles near the banks of the low-water channel, so that the two channel banks become the highest areas on the flood plain. Such banks, or *natural levees,* may actually grow so high that the normal river surface is above the adjacent flood plain. More commonly, however, the alluvium is deposited in shifting point bars as the river swings laterally across the flood plain. Detailed studies of alluvial plains confirm that most of the alluvium is deposited by lateral, rather than vertical, accretion. Vertical accretion on flood plains would eventually build them up so that floods would become less frequent, but the regular periodicity of floods implies that net vertical accretion is slight unless men have cleared the watershed for farming, or otherwise changed the hydraulic geometry of a river.

A river with an alluvial flood plain readily assumes the most favorable channel configuration for the discharge and load that it must move. Rivers that must transport fine-grained sediment in suspension typically have channels that are deep in proportion to their width. Streams that receive large loads of sand and gravel from their headwaters typically evolve wide, shallow channels with a maximum bed surface area. This channel form is most efficient for transporting bed load. If the bed load is excessive, streams assume a *braided* channel pattern, with water flowing in anastomosing, shifting, shallow channels between islets and point bars. During low discharge periods, a braided stream may disappear from the surface entirely, although water permeates the alluvium at a shallow depth. During high discharge, the entire flood plain is awash, and the bed load is dragged forward by the velocity of the water over it.

The rivers of the drier parts of the Great Plains in central North America furnish many good examples of braided channels. Most of these rivers flow down a gentle regional gradient of about ten feet per mile, and many of them lose water by evaporation and infiltration as they leave the mountains and cross the semiarid plains. The local rocks weather to sand and gravel, and mass-wasting imposes heavy loads on the rivers. The American settlers who moved west in the nineteenth century were startled by these wide, flat, often dry river beds that were "too thin to plow, too thick to drink." Loose, water-saturated sand could engulf cattle and wagons. The channels shifted after every rain. Although strange to people from more humid regions, the braided rivers of the Great Plains are nicely adjusted to the work they must do.

Rivers not only transport the sediment supplied by mass-wasting, but they also erode the bedrock over which they flow. The detrital load of a stream provides tools with which the flowing water can abrade solid rock. The force of the current can pry loosened blocks free and pound them together until they break. Chemical reactions with the water can corrode the stream bed. Rivers that do not have their channels lined with alluvium are active agents of erosion. Particularly in the early stages of valley development, as in rugged mountain areas, rivers aggressively deepen and widen their channels into solid bedrock. Canyons thousands of feet deep testify that the erosive force of flowing water is capable of major alteration of landscapes.

How much rock waste do rivers annually move to the sea? The answer varies with the assumptions that are used to answer it. Because of the great variations in discharge and sediment loads, average values may be in error by 50 per cent or more. Bed load is generally assumed to be approximately 10 per cent of the suspended load, but may exceed 50 per cent of the load in braided rivers. Dissolved load is generally somewhat less than the suspended-sediment load, but in the humid, heavily wooded southeastern United States, the dissolved load is 56 per cent of the average total load in transit. The St. Lawrence River carries practically no suspended load, because the Great Lakes act as settling basins for the solid detritus washed in from the headwaters. The dissolved load of the St. Lawrence is 88 per cent of the total load. The Mississippi River, which drains 40 per cent of the conterminous United States, carries about 65 per cent of its load in suspension, about 29 per cent in solution, and about 6 per cent as bed load.

A recent re-evaluation of stream transport in the United States by Sheldon Judson and D. F. Ritter concluded that on the average, each square mile of the United States annually loses 461 tons of weathered rock to the sea. This rate of erosion is sufficient to lower the landscape 2.4 inches per 1,000 years, or one foot in 5,000 years. These rates are slow by human standards, but in the vast perspective of geologic time, entire continents could be washed away hundreds of times over. The average altitude of the continents is 2,700 feet. If erosion were to continue at rates presently typical for the United States, only 13 to 14 million years would be needed to reduce the entire world landscape to sea level. Obviously, as landscapes are lowered, the potential energy of flowing water and the rate of erosion decrease. Nevertheless, unless internal forces of the Earth periodically restored continental altitude, we would live on or in a hydrosphere.

The Concept of the Graded Stream

The term *graded slope* was introduced in the previous chapter to describe a slope that tends by self-regulating processes to maintain itself in the most efficient possible configuration. We are now ready to extend the concept of grade to river systems.

Variables of the Graded River

At least ten variables are involved in the tendency for a river to maintain a graded state. Not all are of equal significance, and some are not within the self-regulatory ability of the river. We divide the variables of the graded state into three classes, *independent, semidependent*, and *dependent.*

Discharge, sediment load, and ultimate base level are the three independent variables in the graded state. The stream has little control over these factors; rather it must adjust to them. Discharge is determined by precipitation and evaporation in the watershed, the permeability of the soil, the amount and type of vegetation, and the area of the watershed. Only the area of the watershed is affected by other changes in the river system. Headward erosion of first-order tributaries can enlarge the watershed area and thereby increase discharge, but even this process is limited because adjacent drainage nets are most probably enlarging, too. Quite early in the erosional development of a landscape, the boundaries of each river's watershed become defined.

Sediment load is also nearly independent of the other variables of stream flow. Many of the same climatic, soil, and biologic factors that determine discharge also determine the amount of sediment that slopes deliver to streams. The type of bedrock is an additional powerful control for sediment load. Some rocks weather quickly to sand-size particles; others produce only silt and clay. Limestone weathering produces mostly a dissolved load, with little solid detritus. Streams erode their channels, and thereby have some self-regulation of their load, but we have seen that most of the load reaches the river "ready-made" by weathering and mass-wasting.

The ultimate base level of erosion is the third independent variable of stream flow. When a stream reaches the sea, it loses its identity. The potential energy of the stream is set by the altitude above sea level at which precipitation falls. Regardless of the discharge, load, or any other variable, a river that rises on a coastal plain only a few hundred feet above sea level will never be a mountain torrent.

The semidependent variables that interact to achieve the graded state include channel width, channel depth, bed roughness, grain size of the sediment load, velocity, and the tendency for a stream to either meander or braid. These are semidependent inasmuch as they are partly determined by the three independent variables, but are also partly capable of mutual self-regulation in a river. The semidependent variables interact in a manner that challenges the intellect of any observer.

Consider the braided habit of some rivers. A braided channel can be defined simply as one with a large width-to-depth ratio. But we have seen that both width and depth are exponential functions of discharge, thus discharge must be a factor in producing braided channels. More significant than discharge is the amount and grain size of the sediment load. S. A. Schumm found that the width-to-depth ratio of rivers on the Great Plains is inversely proportional to the percentage of fine-grained sediment in the load. Sediment grains of sand size and larger travel as bed load; silt and clay sizes (combined for convenience as mud) generally are transported as suspended-sediment load. Schumm in 1960 expressed this relationship between channel shape and sediment grain size by the equation:

$$w/d = kM^{-1.08}$$

where w = channel width, d = channel depth, M = percentage of load in mud sizes, and k = a numerical conversion constant. In many Great Plains rivers the bed load may exceed half of the total load, so the value for M is low and inversely the w/d ratio is high. The braided habit of these rivers is partly an internal adjustment between semidependent variables of width, depth, and sediment grain size, and partly a response to the independent variables of discharge and load.

The meandering habit of many rivers, especially those that flow on alluvium in humid regions, is also related to the width/depth ratio of the channel and sediment grain size. As the suspended-sediment load (mud) increases in proportion to bed load, the width/depth ratio decreases and the channel narrows and deepens. By these interrelated adjustments, more of the energy of the stream is expended against the banks, and less against the deep bottom. The sinuousity of the channel increases, and meanders form.

Free meanders are remarkably regular, and their dimensions are proportional to channel width. The radius of curvature of meanders is usually between two and three times channel width. The wave length of most meanders varies from seven to ten times the channel width. It is the small variation of these and other related ratios that makes the patterns of rillwash across bare ground after a rain look like the patterns of major river channels that are photographed from orbiting satellites. Meandering involves inherent properties of flowing water, as well as size and shape of the channel, erodibility of the stream banks, proportion of suspended and bed load, and probably other factors. In turn, meandering increases the channel length between two points and thus decreases the slope of the stream. Slope influences velocity and sediment-transporting capacity, so meanders not only are affected by other variables of stream flow, but in turn affect those same variables.

Sediment grain size has been repeatedly implicated as a determinant of channel shape and sinuosity. One might think that the grain size is determined only by the nature of the bedrock and the weathering processes in the watershed. As sediment is transported toward the sea, however, the particles are abraded by contact with one another, or partially dissolved by river water. Particle size generally decreases downstream, primarily because of abrasion but partly because of downstream changes in channel shape and bed roughness. As water turbulence decreases downstream, only finer particles remain in suspension. But bed roughness, one of the semidependent variables that causes turbulence, is in part determined by the size of sediment grains that were previously transported and deposited at that part of the river. Thus, grain size of the sediment load and the channel roughness at a given point mutually interact.

Velocity, the remaining semidependent variable, is determined by the equation $Q = wdv$. A change in either discharge, channel width, or channel depth can affect the velocity of flow. Velocity is also affected by the amount and grain size of the load. At one extreme limit, the load carried by a mass of water becomes so great that we call the mixture a mudflow rather than a river. The velocity of a mudflow is much slower than that of a stream, unless the flow is down a very steep slope.

Only one variable of stream dynamics, the downstream slope of the water surface, is regarded as being dependent on all other variables. Slope can be changed only by building up one part of the channel, cutting down another, or changing channel length as by meandering or delta-building. All these changes take time, so slope is usually the final adjustment the stream makes in becoming graded. If slope could change suddenly, it would be mutually semidependent with the variables previously listed. Since it ordinarily cannot do so, it is subject to the influences of all the other known variables.

Progressive Development of Grade

We can visualize the achievement of grade, or the graded state, in a river by first imagining an ungraded river that flows over a tectonic landscape, newly raised from the floor of the sea. Rainfall, and therefore discharge, varies over the new landscape. Water flows downhill along chance hollows and other water-collecting slopes. Valleys are eroded by the water, but the variables of the stream system are wildly out of equilibrium. Half of the potential energy of such a river might be expended in one great vertical waterfall. The load of the stream will be determined by chance landslides anywhere along the banks. Channel width and depth will be restricted by the erodibility of the rocks that are crossed by the channel. The slope of the river will approximate the initial slopes of the landscape.

Rather quickly, the semidependent variables of stream flow interact to form a channel system appropriate to the work at hand. Many of the hydraulic geometry equations for stream flow apply to both graded and ungraded streams. The independent variables of discharge, load, and base level may be the same for a graded and an ungraded river. The critical factor in the achievement of the graded condition is that the stream must flow on "adjustable" materials, so that changes of one variable can produce the appropriate changes in others. Alluvium is the ideal bed for a river, and the establishment of an alluvium-lined channel on a continuous flood plain signals the achievement of grade in that part of a river.

Alluvium not only permits stream channels to adjust toward equilibrium, but it is a powerful absorbent for peak energy inputs into the river system. Discharge may increase by many orders of magnitude during a flood (Fig. 4–2). If the excess energy of the flood is not fully absorbed by the increased load carried into the river from headwater slopes, then alluvium is eroded until the ability of the river to do work is balanced by the work it is doing. Alluvium is analogous to a chemical buffering compound that is added to a solution in great excess to insure that some property of the solution remains constant during a reaction. Alluvium is an excellent buffer for variations in stream energy, for every fragment on the flood plain has some critical energy threshold of transportation. Having been once transported by the river, it will move again when the energy level is high enough.

Grade is established first in the downstream segments of rivers, and is gradually extended upstream. Larger rivers attain grade earlier than smaller streams. A trunk stream may be at grade when its first-order tributaries are still gnawing headward into undissected slopes. A *graded reach*, or graded segment of a river, may form on easily eroded material upstream from a resistant rock barrier. The resistant rock then forms the *local or temporary base level* for the graded reach. As the barrier is slowly eroded, the graded reach remains at grade, for it can adjust easily during the slow lowering of the temporary base level.

Graded reaches of large river systems usually end abruptly downstream at a gorge or canyon through the controlling rock barrier. Some of the most beautiful landscapes of the world are along graded reaches of rivers. The fabled and disputed Vale of Kashmir is one. The upper Cauca Valley of Colombia is another. The "Red Basin" of Ssu-ch'uan (Szechwan) Province in southwestern China is a large-scale example of a graded reach. Four large tributaries of the

Yangtze River meander across easily eroded red shale beds in the basin, 1,000 to 2,000 feet above sea level, before they join and flow as rapids through a series of deep gorges. Perhaps it is the fine climates of these upland broad valleys, or perhaps it is the contrast of their flood plains and meandering rivers with the treacherous gorges by which travelers formerly had to enter, that has given these graded reaches of rivers such reputations for beauty.

Grade as a Thermodynamic Equilibrium

Another approach to the concept of grade is to regard the graded condition of a river from the viewpoint of theoretical thermodynamics. In any steady-state physical system through which matter and energy move, we find a tendency for the least possible work to be done, and also a tendency toward a uniform distribution of work. Nature is inherently conservative in these matters. In a river system, which derives its energy from water flowing downhill, the tendency for minimum work opposes the tendency for a uniform distribution of work.

If all the water of a river were added at the head of a single tributary, the "least-work" profile would be a waterfall straight down to sea level. In humid regions, where rivers gain water downstream, the "least-work" profile is a curve in which the greatest loss of altitude takes place where the discharge is least, near the head of the river. The profile is very steep near the head and almost horizontal near the mouth.

In contrast to the sharply concave theoretical profile of "least work," the theoretical profile for uniformly distributed work has a nearly constant slope. The profile is only slightly concave skyward, because as the rate of doing work increases downstream with discharge, the surface area of the stream bed also increases. The rate of doing work per unit area of stream bed is constant on a river that widens downstream but decreases its slope only gradually.

If two tendencies toward equilibrium oppose each other, the resulting equilibrium is most likely to be some predictable intermediate condition. W. B. Langbein and L. B. Leopold concluded in 1964 from their studies of river dynamics that both the downstream profiles and channel cross sections of rivers approach the equilibrium form predicted from the principles of "least work" and uniform distribution of work, *provided that the channels are in material that is adjustable.* The presence of an alluvial flood plain is again implied as a condition of grade.

The application of thermodynamic principles to the description of graded streams is not simple, and rivers are sufficiently complex so that rigorous "laws of stream flow" may never be written, but it is satisfying to discover that graded rivers with alluvial flood plains apparently meet certain broadly defined conditions of equilibrium that are typical of the steady-state open systems maintained during carefully controlled laboratory experiments.

Summary of Grade

The most concise definition of a graded stream was carefully phrased by J. H. Mackin in an essay on the subject written in 1948. The hydraulic geometry of Leopold and Maddock required only a slight transposition of phrases in Mackin's definition. It is quoted here, as modified by Leopold and Maddock, as a summary of the concept of the graded river:

> A graded river is one in which, over a period of years, slope and channel characteristics are delicately adjusted to provide, with available discharge, just the velocity required for the transportation of the load supplied from the drainage basin. The graded stream is a system in equilibrium; its diagnostic characteristic is that any change in any of the controlling factors will cause a displacement of the equilibrium in a direction that will tend to absorb the effect of the change.

A few additional comments must be made about the concept of grade. It is a condition, not an altitude or a certain slope angle. It develops first near the mouths of rivers and gradually extends headward. Erosional lowering of the land continues for a long time after grade is achieved, because as long as rivers carry sediment to the sea, they continue to lower the landscape over which they flow. A graded river is in a steady state only with regard to short-term changes. Over a time scale of millions of years, typical of the time intervals in which landscapes evolve, the potential energy of an undisturbed river system gradually approaches zero, and the rate of change of the system also decreases. The river remains at grade, but the characteristics of the graded condition change with time. The significance of the concept of grade to the life history of regional landscapes is emphasized in Chapter 5.

Water in Dry Regions

Water is the greatest agent of geomorphic change. When it is in deficient or irregular supply, the landscape reflects the deficiency in diagnostic ways. The most distinctive features of dry landscapes are related to the work of flowing water, because weathering and mass-wasting in dry regions and humid regions differ only in degree. The work of flowing water in dry regions nicely illustrates some of the broader principles of graded slopes and channel formation that have been described on previous pages. The remarkable development of one landform, the *pediment,* under dry climates is particularly worthy of our attention. Furthermore, water is the key to economic development of dry lands, and an understanding of its work has many practical applications.

Dry Climates

By definition, in a dry climate potential evaporation from soil and vegetation exceeds the mean annual precipitation. The temperature of the region is not important to the definition except as it helps determine potential evaporation. Dry climates are divided into two degrees of intensity, semiarid or steppe, and arid or desert. The semiarid climate is separated from humid climates by a ratio of precipitation to evaporation that is less than one. The dividing line between desert and steppe is arbitrarily taken as one-half the amount of precipitation that locally separates steppe from humid conditions. No absolute annual amounts of precipitation can be included in these definitions, for as the average temperature increases, the amount of precipitation needed to exceed evaporation also increases. In the United States, the minimum annual precipitation required for a humid climate increases southward from about 20 inches in North Dakota to about 30 inches in Texas. Less than 10 inches of annual precipitation will produce a desert in almost any temperature range.

Dry climates cover more land area than any other climatic type (Fig. 4–4). About 26 per cent of the world's land is dry, an impressive proportion when one considers the abundance of water on the surface of the Earth. In low latitudes, two belts of aridity coincide with the subtropical anticyclonic belts of high atmospheric pressure about 15° to 30° north and south of the equator. These deserts are determined by the planetary atmospheric circulation pattern. They are fringed by relatively narrow belts of semiarid transitional climates.

In middle latitudes, dry climates are localized in the interior of the large continents. Temperature and evaporation are not so high in these regions as in the subtropics. Some precipitation may fall as snow. Mid-latitude dry climates are characterized by relatively large areas of semiaridity around smaller cores of true desert. Small areas of dry climate are also found on the west sides of the continents where cold ocean currents flow offshore, and on the downwind, or rain shadow, side of mountain ranges. The polar deserts are excluded from this account because they are largely ice-covered.

Geomorphic Processes in Dry Climates

The geomorphic significance of dry climates begins with the vegetation. Semiarid climates usually support a sparse grassland, or steppe, vegetation. In the United States, the boundary between humid and semiarid climates is approximated by the transition westward from medium-height grasses with a continuous turf or sod in the humid regions to short, shallow-rooted bunch grasses on otherwise bare ground in the semiarid regions. In arid regions, even the bunch grasses disappear, and the vegetation is, at best, widely spaced shrubs and salt-tolerant bushes.

Wind assumes importance as an agent of erosion and transportation in

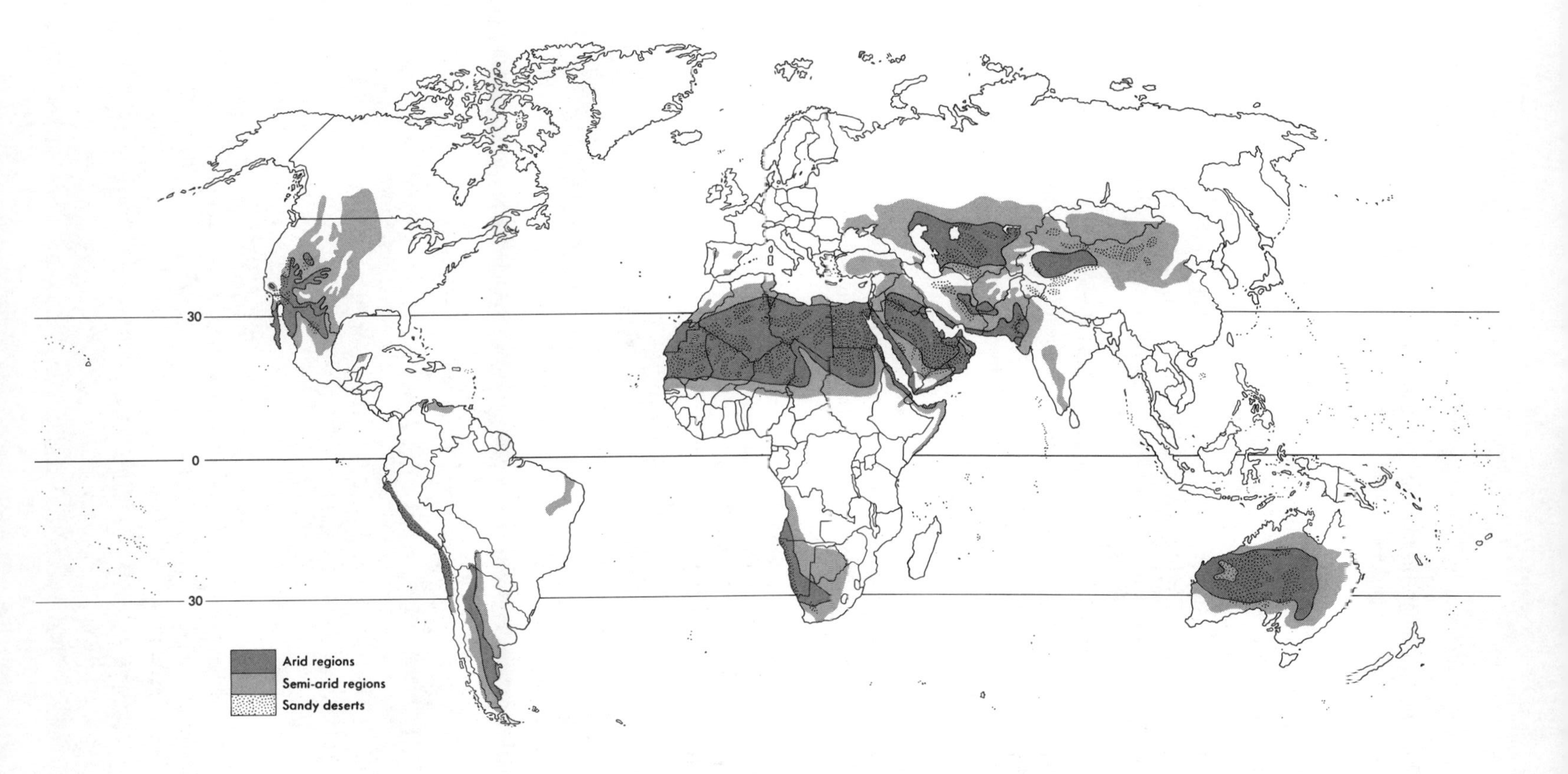

FIGURE 4–4 *Low latitude and mid-latitude arid and semiarid regions of the world. Darkest areas are arid, lighter shading is semiarid. Sandy deserts are stippled. Polar regions, not shown here, are also among the driest places on Earth, both because of low precipitation and lack of H_2O in liquid form.*

deserts. Dust and sand can be swirled up from bare ground and removed from the area in the erosional process called *deflation*. Generally, sand is moved by wind in low bouncing arcs no more than a few feet above ground level, but dust particles may be carried thousands of feet into the air and across continents. At the height of the "dirty thirties," a time of extensive drought in North America, eastern American cities were made twilight-dark at midday by dust blown east from the Great Plains, 2,000 miles away. Wind-transported dust eventually falls in the sea or forms a thin blanket-like deposit over the downwind landscape. Wind-blown sand is quickly trapped by obstructions and accumulates as *dunes* a short distance downwind from the source area (Fig. 4–5).

Desert rains are usually of high intensity, short duration, and local extent. Convective thunderstorms, or *cloudbursts,* over desert mountains bring most of the rain. Bare, sun-dried ground promotes rapid runoff and *flash floods.* Light surface runoff may travel only a short distance before it is absorbed into the dry, cracked desert soil. Heavier surface runoff moves as a *sheetflood* of muddy water, too loaded with suspended sediment to erode more than incipient shallow channels that fill as rapidly as they form.

Precipitation in semiarid or steppe regions is also of brief duration and high intensity, but the frequency is somewhat greater than in deserts. Most semiarid

FIGURE 4–5 *Great Sand Dunes National Monument, Colorado. These dunes reach heights of 700 feet. The sand is blown northeastward against the base of a mountain range by strong winds across the dry San Luis Valley in the distance. (Courtesy U.S. Department of Interior, National Park Service.)*

regions have distinct seasonal patterns of precipitation, so that for at least part of the year the ground may be covered with vegetation. Enough excess precipitation falls to provide soil infiltration and runoff to rivers for part of the year. In mid-latitude steppes, summer is the time of most precipitation, but evaporation is also most effective then. Snowfall and rain during the seasons of lower evaporation are necessary to recharge the ground moisture sufficiently to permit runoff.

With a deficiency of water, dry soils do not develop strong diagnostic horizons except for salt crusts or concretionary layers. The Mollisols of the temperate grasslands (Table 2–1) give way to the Vertisols, Aridisols, or Entisols of semiarid and arid regions. The definitive climatic criterion—precipitation less than evaporation—means that if soil moisture is available by lateral infiltration, as for instance in a desert basin flanked by more humid mountain slopes, the water is drawn upward through the soil by capillarity and evaporates either in the soil profile or at the ground surface. Dissolved mineral matter is precipitated when the soil water evaporates. Calcium carbonate (caliche), gypsum, or alkali may accumulate in desert soils in this manner. Except by upward movement of ground water, there is little opportunity for weathered or soluble material to move vertically in a desert soil to form mineral horizons, and no humus layer develops under the sparse vegetation.

Steppe soils are better formed than desert soils, but nevertheless they have poorly developed horizons and are thinner than soils in humid regions. In wet years, grass and shrubs may form a nearly continuous ground cover on steppe lands. Evaporation is rarely so extreme that the more soluble salts in the soil water are precipitated, but relatively insoluble calcium carbonate may form nodules or layers in the soil profile from either downward or upward movement of soil water.

Weathering in dry regions occurs by the same processes that operate in humid climates, but the general rate of arid-climate weathering is very slow. The rare rainfalls and night dew apparently provide the moisture for hydrolysis and hydration. The puzzling fractured stones of hot deserts were described in Chapter 2. Mechanical weathering becomes relatively important in dry regions as compared to chemical weathering, but both classes of weathering are much slower, and many researchers believe that even in the driest deserts, chemical decomposition involving water dominates over mechanical disintegration.

Soil creep is almost unknown in deserts, for the soil is not bound together by a turf of interconnected roots. Other varieties of surface creep are common, however. Rock creep and rockslide are even more obvious without a concealing blanket of sod. Taluses of coarse sliderock mantle many desert slopes. Summit convexity is almost absent from desert slopes. The typical desert profile has a cliff and a talus that rise above a concave wash slope. Rock structures stand out boldly.

Slopes in semiarid regions reflect the more complete vegetation cover and more intensive soil-forming processes that near-adequate precipitation provides. Soil creep maintains summit convexities wherever vegetation and soil permeability inhibit sheetwash. In mid-latitude and high-altitude steppes, frost wedging in the winter probably aids other causes of creep.

Desert Landforms

Stream channels in arid regions commonly have rectangular cross sections with nearly vertical banks and flat alluvial beds. The contrast with upward-flaring trapezoidal or arc-shaped cross sections of humid-region channels is due to the highly intermittent nature of stream flow in dry climates, the generally high proportion of the stream load transported as bed load, and the tendency for dry soils to crack into vertical columns. Every dry region has a distinctive name for steep-sided, flat-floored stream channels. In Latin America and the Spanish-American southwestern United States they are called *arroyos*. In parts of the Sahara where French is spoken they are *ouadis*, an obvious derivation from the original Arabic term, *wadi*. Other descriptive terms are *wash* and *draw*.

During runoff, the entire floor of an arroyo will be awash. The banks are undermined, and debris that breaks from the channel walls along joints in bedrock or dessication cracks in unconsolidated material falls directly into the stream and is swept away. The downvalley slope of arroyos is steep as a consequence of their excessive sediment load.

Commonly, steep gradients of mountain streams decrease abruptly where the streams discharge from the mountain front into a tectonic basin or the valley of a more gently sloping master stream. The alluvium and mudflow debris dropped by such a stream at the point of slope reduction builds a conical mound with its apex at the mouth of the canyon in the mountain front. The slope down all radii of such an *alluvial cone*, or *alluvial fan*, is approximately the same, for as soon as one radius of the fan is built up by deposition, the channel spills over to an adjacent position. The size and symmetry of alluvial fans are impressive features of mountainous desert landscapes (Fig. 4–6). Contours on alluvial fans are commonly arcs of concentric circles centered at the canyon mouth. The profiles of alluvial fans are concave skyward, typical of wash slopes. The surface of an alluvial fan is the downstream continuation of the steep valley-floor slope of a canyon or an arroyo upstream from the apex of the fan. The surfaces of alluvial fans, perfect examples of water-spreading wash slopes (Fig. 3–9, quadrant IV), are ideal forms for the dispersal of peak discharges and sediment loads from mountain streams during flash floods. The surface of an alluvial fan may be covered with braided channels, or it may have only a few radial channels that successively shift across the fan.

The largest alluvial fans build into the arid intermontane basins that form local base levels for erosion of the adjacent mountain slopes. Many of the intermontane basins of the arid southwestern United States have tectonically subsided as the adjacent mountain blocks have risen. In these closed basins allu-

FIGURE 4–6 *An alluvial fan in Death Valley, California. (Courtesy John S. Shelton.)*

vium many thousands of feet thick has accumulated. The Spanish name *bolson* (meaning purse) is an apt name for the mountain basin that stores the weathered debris from the mountains that encircle it (Fig. 4–7). The surface of such a basin consists of coalescing alluvial fans, or a *bajada.* At the lowest altitude on the bajada, an intermittent *playa lake* may form after a heavy runoff. When the lake evaporates, a very flat *playa,* encrusted with salt or dried mud, marks the local base level of erosion.

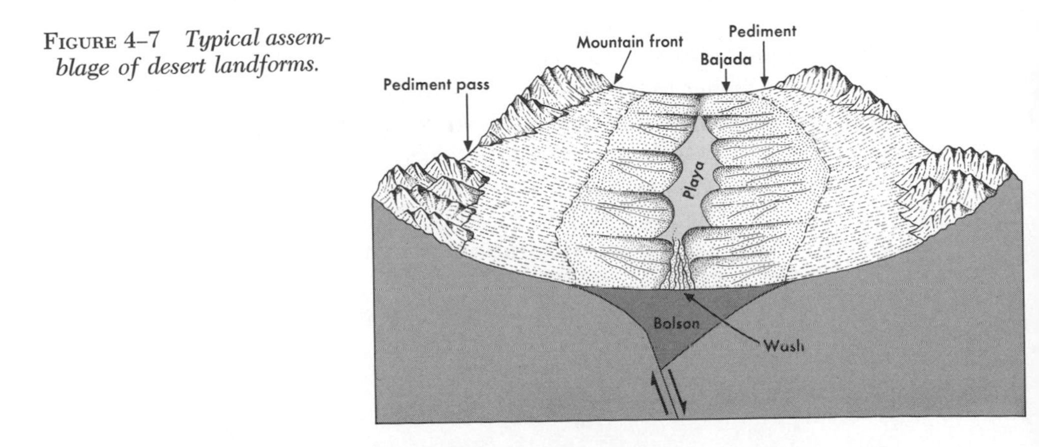

FIGURE 4–7 *Typical assemblage of desert landforms.*

A curious but significant feature of landforms in arid regions is that the playa is a rising base level. As the bolson is filled, the alluvial fans constantly thicken and engulf the mountain fronts that feed them. Because of the geometry of a bolson, the surface area increases as the level of the alluvial fill rises, so that even a constant rate of mountain weathering and erosion will result in a decelerating rise of local base level. The erosional consumption of the mountains with time and the loss of potential energy as the mountains are lowered, further insure a decelerating rate of bolson filling. Arid regions with interior drainage are fundamentally different from all other regions in that the base level of erosion is local and rising, instead of ultimate and approximately stable. The consequences of this difference will be fully analyzed in the next chapter.

Pediments

The most significant component of dry landscapes is the *pediment*. The name refers to the eroded wash slopes that rise toward the base of mountains in dry regions. From a distance, a desert mountain seems to stand at the crest of a low-pitched roof, and the term "pediment" was borrowed from classical architecture where it refers to the triangular end or gable of a building.

As first used by G. K. Gilbert in 1880, the term pediment was applied to the surfaces of the alluvial fans that encircle desert mountains. As exploration of the arid southwestern United States continued, it was realized that large areas of the wash slopes around mountains are not built of alluvium, but are eroded across solid rock. W J McGee in 1897 published a graphic description of the pediments in the Sonoran Desert of Arizona and northwestern Mexico, in which he wrote of his surprise that, "the horseshoes beat on planed granite or schist or other hard rocks in traversing plains 3 to 5 miles from mountains rising sharply from the same plains without intervening foothills." He concluded that fully half of the intermontane slopes of the district were cut across bedrock and were either bare or veneered with alluvium so thin that it could be moved by a single sheetflood.

Following McGee, Kirk Bryan in 1922 formally gave the name "pediment" to a surface formed at the foot of an arid or semiarid mountain by the erosion and deposition of ephemeral streams. He emphasized in his definition that pediments are slopes of transportation, usually covered with a veneer of alluvium in transit from higher to lower levels. It is this emphasis on the pediment as a slope of transportation uniquely adapted to weathering and erosion in dry climates that validates its inclusion in a general book on geomorphology. We have in the study of pediments an opportunity to reconsider the concept of grade, and the relation of sediment in transport to the work of flowing water, in the context of a climate different from the humid one in which most of the world's people live.

Pediments have a surface form indistinguishable from alluvial fans. Both are water-spreading wash slopes (Fig. 3–9, quadrant IV) uniquely adapted to the bare ground and rapid runoff of flash floods. When a flash flood emerges from a canyon onto the lower mountain slopes it spreads laterally and thins to a *sheetflood.* Loose sediment is quickly swept up by the flood, and even if it was not fully loaded as it rushed down the canyon, the water on the lower slope is quickly loaded to capacity. Infiltration, evaporation, and lateral spreading reduce the discharge downslope and further burden the flowing water with mud and rock debris. A sheetflood advances at race-horse speed as a wall of water a foot or more in height, surging forward in lobes around obstacles. The suspended load may make the water so dense that large boulders bounce along like corks half afloat in the muddy water.

If the current of a sheetflood is restricted by a clump of bushes or other obstruction, some of the load is dropped. Relieved of its load, that water surges forward and immediately scours a pit in the alluvial veneer on the pediment. In this manner, a single sheetflood may spread several miles down a pediment before it concentrates to a mudflow and stops. Presumably, a sheetflood is able to move intermittently all of the alluvium on the pediment in addition to the load it derives from erosion within the mountain headwaters. The entire alluvial veneer is never in motion at one instant, but by repeated cutting and filling, the sediment is moved downslope.

Sheetflood is generally described as the opposite of streamflood, or channeled flow. Like the sheetwash or rillwash of concave humid-region slopes, but on much greater scale, sheetfloods are so burdened with their load that they have no kinetic energy available to erode channels. In humid regions, discharge and transporting efficiency increase downslope and the water gathers into channels; in dry regions, the discharge decreases downslope and evaporates or is absorbed by the load.

As dry mountain slopes erode, pediments form at their bases. Tongue-like projections of the pediments may extend up into the mountain mass along major valleys. Some dry mountain ranges have broad *pediment passes* extending through them, the result of the headward expansion and intersection of pediments from opposite slopes of the range. Pediments widen by the lateral migration of rills and ephemeral channels that periodically impinge on the mountain front and undercut taluses and cliffs. They also expand by the normal mass-wasting retreat of cliffs and taluses, and by abrasion of the rock-cut pediment surface by sheetfloods. The relative importance of the three processes seems to vary from region to region. Some pediments have numerous shallow arroyos entrenching their surface and are thought to be eroded by the lateral shifting of the channels. Other pediments lack evidence of any permanently channeled flow. Analysis of these details is difficult because all erosion is slow in dry regions, and some pediments probably formed under a different climate and are now being slowly dissected. Thus, an analysis of present processes may not give a true impression of the origin of a pediment.

As pediments expand into the mountain slopes that feed them water and rock debris, the lower part of the pediments in turn may be progressively buried by the rising bajada alluvium (Fig. 4–7). For this reason, pediments in truly arid regions, where the local base level is a playa, are generally narrow

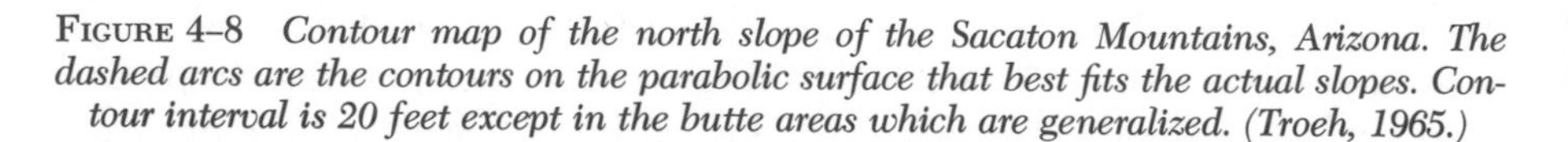

FIGURE 4–8 *Contour map of the north slope of the Sacaton Mountains, Arizona. The dashed arcs are the contours on the parabolic surface that best fits the actual slopes. Contour interval is 20 feet except in the butte areas which are generalized. (Troeh, 1965.)*

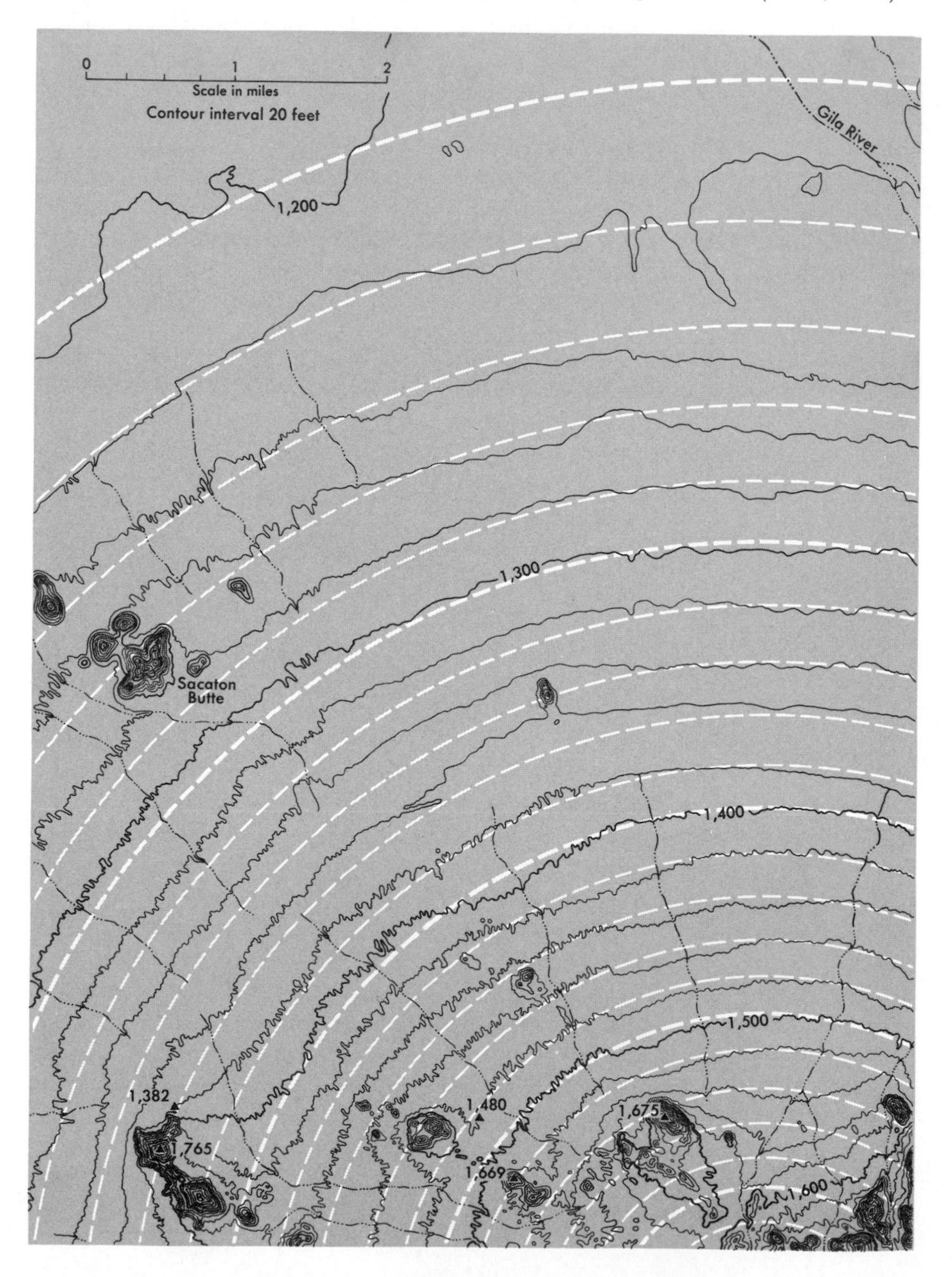

rock-cut fringes between mountain fronts and the bajada. The pediments plunge downslope beneath the thickening alluvial fill of the bolson. By those who need precise definitions, the smoothly graded wash slope at such a mountain front would be called a pediment in its upper part, where the entire thickness of alluvium is in transit, and a bajada in its lower part, where the alluvium has permanently come to rest in the bolson. The boundary between pediment and bajada is then defined by the thickness of alluvium that can be periodically reworked by sheetflood or arroyo cutting and filling.

The most extensive pediments are found in regions that have sufficient precipitation to allow at least occasional runoff to the sea. Under these conditions, which are technically semiarid rather than arid, weathered sediment can be transported downhill to the ultimate base level and be completely removed from the subaerial landscape, instead of filling bolsons and progressively burying the lower edges of developing pediments. The Sonoran Desert in Arizona and the neighboring Mexican State of Sonora form the classic region of pediment development in North America. The region has steep, alluvium-floored river channels graded to sea level, although water rarely flows in them. The entire Sonoran landscape may be relict from a time of greater precipitation.

The interior plateaus of large parts of Africa are reported to be great, gently sloping pediments, and large parts of the dry continent, Australia, are also pediment prone. The arid interior of Asia is less well known than other dry regions, but there too, pediments can be predicted to be among the dominant landforms. Indeed, considering that 26 per cent of the land area is arid or semiarid, the pediment may be the most common erosional landform on Earth.

To illustrate the form of a pediment, F. R. Troeh applied his technique of slope analysis (Chapter 3, Fig. 3–9) to a well-known pediment on the north slope of the Sacaton Mountains in southern Arizona. Fig. 4–8 is a portion of a contour map of the pediment with dashed arcs superimposed to show the best-fitting surface that Troeh could generate by rotating a parabolic segment around a vertical central axis. The altitude of any point Z on the surface of the pediment is given by the equation:

$$Z = 1708 \text{ ft} - (102.8 \text{ ft/mile})\, R + (5.06 \text{ ft/mile}^2) R^2$$

where R is the horizontal radial distance of point Z from the axis of rotation. The equation is in the form of a general quadratic, but is written in the reverse of the usual order as:

$$Z = c + bR + aR^2$$

The altitude at the apex of the surface is given by the constant c, the initial slope by b (a negative slope, here), and the rate of change of slope radially outward from the origin is $2a$. The slope at any point Z is given by the derivative:

$$\frac{dZ}{dR} = b + 2aR$$

Mathematically, the equation describes the pediment in Fig. 4–8 as a parabolic cone, centered at a hypothetical point of origin 1,708 feet above sea level, that has an initial slope of 102.8 feet per mile (downhill) and a rate of decrease of slope of $+2(5.06)$ or $+10.12$ feet/mile over each radial mile. The equation fits the pediment almost perfectly, and departs from the land surface only where the unconsumed mountains stand at the crest of the pediment, and at the extreme northern edge of the map where the pediment ends at the flood plain of the Gila River. Both in the mountains and on the flood plain, other types of erosion and transport operate. For a radial distance of at least seven miles across both exposed rock and alluvium, the pediment is an ideal water-spreading wash slope, perfectly graded to move large but intermittent masses of water and weathered rock from the Sacaton Mountains downhill to the Gila River. The continuity of slope indicates the continuity of process, and the pediment is an ideal example of a slope of transportation that remains graded as the landscape is gradually lowered by erosion.

5

Life history of landscapes

Deductive Geomorphology

The preceding four chapters have been concerned with the processes that change landscapes, the rates at which the processes work, and the observational evidence that landscapes differ from place to place because of the nature, intensity, and duration of the processes of change. We have reasoned from specific examples toward statements of general principles. This is the *inductive* approach, and it is a vital part of the system of logical thinking we call science.

The other side of the logical coin of science is the *deductive* approach. Here, we reason from general principles toward analysis of specific problems. The deductive approach to geomorphology primarily concerns the changes of landscapes through time. We cannot watch a landscape evolve, even though we have abundant reasons to think that it does. In our deduction, we apply the principles we have learned from studying many places, each at a brief interval of time, to the prediction of events at a single place, during many successive time intervals. We wish to make a motion picture of a landscape evolving, but our source material is a series of still photographs of many different landscapes.

Biologists or philosophers might argue that a landscape is nonliving and therefore has no "life history." But in the same sense that we commonly speak of the life of an automobile, an electric light bulb, or a book, we use life history here for the events that occur during the recognizable duration of a landscape. We must specify the initial form and time of origin of a landscape, and we must specify the conditions of its ultimate destruction. Between these limits, it has a life history to be deduced.

The deductive study of landscapes has been severely criticized for having exceeded the restraints of observational evidence. It is true that W. M. Davis and others in the first decades of this century pushed the deductive system of "explanatory description" well beyond what had been proved experimentally. The great danger of deductive reasoning is that if the general principles are wrong or incorrectly applied, even the most careful logical procedures will inevitably lead to erroneous final deductions. Suppose, for instance, that a general principle had been established through repeated observations in humid regions (it has not) that hillside slopes decrease in steepness through time. If this principle were incorrectly used to predict the future shape of a desert slope, the deduced shape would have little relation to real land-forms, for different processes of mass-wasting dominate in arid and in humid climates.

In recent decades, we have discovered that the repeated glaciations of lands in the middle and high latitudes were but one aspect of repeated general changes in climate during the last 1 to 2 million years. We do not know the causes of these climatic changes, but we recognize their effects on landscapes. Some geomorphologists feel that we should not attempt to deduce the advanced stage of landscape evolution under a certain climatic condition, because long before that advanced stage is reached, new climatic conditions will be imposed that will alter the course of landscape development. They forget that we recognize the effects of climatic change on landscapes only because the landscapes are not what we predict should have formed under the conditions we observe at the present.

In this chapter, the sequential evolution of landscapes under conditions of humid climate and abundant flowing rivers will be given first importance. We will do so even though dry climates cover a larger land area than any other single climatic type. Figure 1–3 is adequate evidence that more water annually falls on the land than evaporates from it, and that the "normal" or "typical" landscape evolves under conditions of excess water runoff. Subsequently, variations on the theme of landscape evolution by flowing water will be suggested. As in a musical composition, the complexities of the variations cannot be appreciated until the theme is known.

If in some people's minds the deductive approach to geomorphology is less precise, more intuitive, and less "scientific" than the inductive method, it still has a vital role in man's understanding of his environment. In attempting to reconstruct future and past conditions, we comprehend the immensity of geologic time and the infinite patience with which natural changes are wrought. We still have the needs of the ancient philosophers who meditated on the dubious durability of hills and thereby found strength for themselves. There is a pleasure in understanding for its own sake.

Proofs That Landscapes Evolve Sequentially

An alternate expression for the life history of a landscape is *sequential evolution.* The latter term implies that one set of landforms evolves gradually into another in a unidirectional sequence unless tectonic or climatic events intervene. There are at least five experimental or observational proofs that landscapes evolve in predictable sequences of forms. Some are more rigorous than others, but all are useful.

1. *Small-scale experiments* on landscape evolution can be conducted on sand tables or other models. A groove scratched in a sloping sand surface will conduct water, and as it does so, will change its shape to accommodate the discharge. A mound of fine sand under a spray mist will develop drainage networks down its slopes. By careful choice of materials and experimental conditions, many landforms can be reproduced in miniature in a laboratory. It would seem that scale-model experiments provide the best proof that landscapes evolve, and so they would, if they really duplicated nature.

The difficulty with all scale models is that changes of dimensions in length, mass, and time do not alter the intrinsic properties of matter. For example, water in a model channel a few inches deep has the same density and viscosity as water in a real river; therefore turbulence in the water due to these properties will be grossly out of scale in the model. Furthermore, as the linear dimensions of an object are decreased, the volume decreases as the cube of the length, but the surface area decreases only as the square of the length. Thus, small particles have much larger surface areas in proportion to their masses than large particles of the same material. Surface effects can cause very fine-grained wetted particles in a model landscape to cling strongly together in a fashion totally inappropriate to the sand or gravel of the real landscape that they are meant to represent in scale. Finally, physical constants such as gravitation cannot be scaled down. An experimental river channel on a sand table proves how river channels form on sand tables, but very little more. Small-scale experiments are useful only to study the intermediate stages in evolutionary sequences that are known to occur from other kinds of evidence, but to that extent, they prove that landscapes evolve.

2. *Real landscapes evolving under accelerated conditions* show us what sequences to expect when changes are too slow to observe, as they usually are.

Complex drainage networks may evolve on tidal mud flats during the few hours of each low tide. After volcanic ash falls have completely obliterated pre-existing landscapes, new drainage systems have been initiated, which then have been observed to develop over a period of months or years. Everyone has seen gullies erode in the artificial fill at construction sites. Larger gullies, up to 100 feet deep and several miles long, have been known to develop in less than a century after land was cleared for agriculture.

Accelerated erosion due to natural or man-made catastrophe is good proof that landscapes evolve, for the changes are full-scale except in time. Unfortunately, the "accident" that induces accelerated change commonly affects only one or a few processes of change. Accelerated erosion due to forest clearing and farming is caused by more rapid and concentrated runoff, but soil formation is not speeded up correspondingly, and mass-wasting may change from soil creep to slumping and earth flow. Thus, the resulting landforms are not those that would have formed if all processes of change had been accelerated proportionally. The results of accelerated erosion are instructive to watch and measure, but they cannot be directly extrapolated to interpret landscape evolution on the scale of geologic time.

3. *Playfair's Law of accordant junctions* contains a third proof that landscapes evolve sequentially. In his *Illustrations of the Huttonian Theory* published in 1802, John Playfair wrote the following sentence, which has few rivals for clarity and style and has become known as a "law," although scientists ordinarily frown on dogma.

> Every river appears to consist of a main trunk, fed from a variety of branches, each running in a valley proportioned to its size, and all of them together forming a system of valleys, communicating with one another, and having such a nice adjustment of their declivities, that none of them join the principal valley, either on too high or too low a level; a circumstance which would be infinitely improbable, if each of these valleys were not the work of the stream that flows in it.

We can recognize that Playfair's statement is not the rigorous proof of a natural law, but only an observation about a highly probable condition. If valleys did not evolve by the work of the streams that flow in them, then the "nice adjustment of their declivities" would be a highly improbable state. Having considered the hydraulic geometry of streams and the concept of graded rivers, the average reader of this book is better equipped to appreciate the significance of Playfair's words than the greatest scholars of 1802.

4. *Rivers consume their own source of energy.* As long as a river is carrying sediment to the sea, it is lowering the landscape that provides the gravity potential for flow. Processes of weathering, mass-wasting, and erosion that depend on the potential energy of position constantly decrease in effectiveness as altitude decreases. Therefore, the entire landscape must continuously change as it is lowered. There can be no long-term steady state in a physical

system of declining energy supply. Graded rivers are capable of adjusting to changes from year to year, but they also gradually adjust to loss of potential energy. A graded river remains at grade while a landscape is lowered, but the conditions of grade continuously change.

The sediment in transport by rivers is a powerful proof that landscapes are evolving. So is the alluvium that is temporarily left behind as *river terraces* during valley deepening. Alluvial terraces prove by their surface features and internal composition that they were once part of a river's flood plain. If they are not now reached by floods, either the valley has been deepened since they formed, or there has been some diversion of water from the river. This proof of valley erosion is another of John Playfair's contributions. Eventually, terrace alluvium must resume its trip to the sea, when weathering and mass-wasting again deliver it to the river that left it behind.

5. *Landscapes from many different areas can be arranged in a single series.* This is not a rigorous logical proof that any single landscape will evolve in the path of the series, but it is still the best practical demonstration of landscape evolution. Earlier it was noted that we want to make a motion picture of an evolving landscape, but our source material is a series of still photographs of many landscapes. The fact that we can make that motion picture is the most convincing evidence of its validity. We can assemble any group of landscape photographs into subgroups of those that look most alike, then within the subgroups we can further arrange the photographs so that transitional forms are closest to adjacent subgroups, making due allowance for regional differences of climates and rock types.

The ability to arrange landscapes in some order does not tell us in which direction the order proceeds. Even if we make our motion picture, we do not know which way to run the film. For the establishment of a unidirectional series, we return to the preceding proof of landscape evolution, that rivers carry sediment. If landscapes form a developmental series, it must be in the direction of larger valleys and smaller residual hills, for weathered rock is constantly being removed to the sea.

The practical fact that landscapes can be sequentially arranged has a theoretical basis in thermodynamic principles. W. B. Langbein and L. B. Leopold, in their 1964 analysis (described in Chapter 4) of the graded state of rivers by the combined principles of least work and equal distribution of work, made the significant inference that as rivers tend toward equilibrium between the two opposing tendencies, "deviations in time will approximate those which can be observed in an ensemble of places." Some conditions of landscapes are more probable than others, and nature favors tendencies toward the probable. The two most probable changes in landscapes are, first, that the rivers will become graded, and second, that their potential energy will decline through time.

It may seem a waste of time and intelligence to outline these proofs of landscape evolution, because they are all quite apparent. You may not realize that only 150 years ago the concept of slow, orderly development of landscapes under the same conditions that operate today was a dangerous challenge to the established religious and philosophical order. The acceptance of Hutton's concept of uniformity and continuity of process and change, and the corollary concept of the enormity of geologic time, set the intellectual stage for the theory of organic evolution in the mid-nineteenth century.*

Initial Landscapes

If landscapes evolve sequentially, there must be original or uneroded stages that can be described. The favorite initial stage postulated by deductive geomorphologists is the former floor of a shallow sea, newly emerged above sea level. The assumptions are that the sea floor is smooth and featureless, and that it emerges quickly relative to erosional processes. Figure 5–1 suggests such a landscape, with scale omitted. The figure could represent 1 or 1,000 square miles.

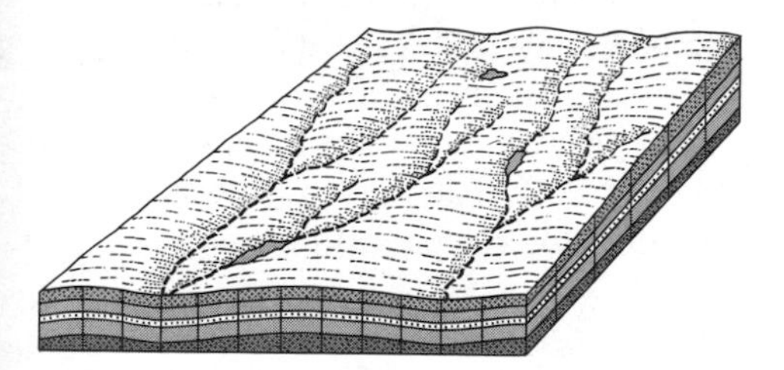

FIGURE 5–1 *A hypothetical initial landscape, newly emerged above sea level. The initial surface is conformable to the sedimentary beds that underlie it. Drainage is consequent on the surface irregularities. (Redrawn from Cotton, 1942.)*

Emergence of the smooth, depositional surface of a shallow sea floor could result from either a lowering of sea level or uplift of the land. If the former, then the only slope, if any, on the new land is the depositional slope of the former ocean floor. If the latter, then the initial slope might be the former slope of the ocean floor, or the slope might be increased or decreased by differential tilting during uplift.

Tectonic uplift is by far the most probable cause for new land to emerge from the sea. Although sea level has fluctuated through a range of 300 to 400 feet because of the repeated expansion and shrinkage of glaciers (Fig. 1–3), these fluctuations have been of too brief duration to affect regional landscapes. (The effect of sea level changes on coasts is described in more detail in Chapter 6.) Tectonic uplifts are divided into two types: *orogenic,* or mountain-building, movements; and *epeirogenic,* or broadly regional, movements that do not intensely deform the rocks.

We can measure some rates of orogenic uplift directly from historic records or prehistoric evidence. Parts of coastal Alaska were uplifted as much as 33 feet as a result of the 1964 earthquake. In New Zealand, where a full-scale orogeny is apparently in progress, some areas are reported to be rising 4 to 35 feet per

* D. L. Eicher's book on *Geologic Time* in this series discusses these interesting matters at greater length.

1,000 years. Rising anticlines (upward-arched rocks) have reversed the direction of drainage across some sloping fields in less than 100 years since the land was cleared for farming.

Precise surveys near Los Angeles show that some mountain masses there are rising 13 to 20 feet per 1,000 years (4–6 mm/year: enough to seriously affect sewers, water pipes, and other carefully graded engineering works). California, like many other areas around the Pacific Ocean, is strongly orogenic. A good estimate for rapid orogenic uplift is about 30 feet per 1,000 years.

Epeirogenic uplift is probably much slower than orogenic uplift. Regions the size of the central United States are known to have been under the sea repeatedly during geologic time, and to have been above sea level and eroding at other times, but the total vertical range of such epeirogenic movements may be only a few thousand feet, and may take millions of years to accomplish. One authority estimates epeirogenic uplift at a few feet per 1,000 years, perhaps 10 per cent as fast as orogenic movements.

The comparative rates of orogeny, epeirogeny, and regional erosion help us establish the likelihood of a landscape rising fresh from the sea, like Venus on a clam shell. The most recent estimates for regional lowering of the United States by erosion (Chapter 4) suggest a rate of 0.2 feet per 1,000 years (1 foot in 5,000 years). Mountainous areas erode much faster, for slopes are steeper, climates are more severe, and the potential energy is much greater. The most rapid rate of regional lowering yet measured is in the central Himalaya Mountains, where the suspended-sediment load alone in one large river indicates a regional erosion rate of about 3.3 feet per 1,000 years. A good estimate of mountain lowering by erosion is about 3 feet per 1,000 years.

The estimated rates of tectonic uplift and regional erosion are summarized in Table 5–1. They suggest that orogenies raise mountain ranges up to ten times

Table 5–1

Comparative Rates of Uplift and Erosion*

	Mountains (orogeny)	**Continental areas (epeirogeny)**
Uplift	30	2–3
Erosion	3	0.2

*Figures given are feet per 1,000 years.

as fast as the most active rivers could erode them down, just as epeirogenic uplift might gently raise a continent about ten times as fast as flowing streams could lower it. There is no point in comparing mountain erosion to epeirogenic uplift, for without the great altitude of mountains, streams cannot work that fast.

All these estimates mean that if tectonic movements raise new land above the sea, the land is likely to increase in area and altitude faster than erosion can destroy it. After the tectonic uplift ceases, erosion must continue for a considerable time before the landscape is consumed. In the previous chapter, it was noted that the present rate of erosion of the United States would reduce an average continent to sea level in 13 to 14 million years. Because the rate of erosion decreases as the potential energy of rivers declines, the actual time would be much longer. The ultimate reduction of a land to sea level will never be accomplished by rivers alone, for the potential energy of rivers approaches zero as a landscape is lowered toward base level. Sea level is the limit of subaerial erosion, and as such it is approached, but not reached.

Tectonic movements can lower land as well as raise it. There is abundant evidence in sedimentary rocks that former landscapes, often with a weathered regolith, have been submerged and buried under new marine sediments, which in time become lithified, uplifted, and exposed by erosion. The irregular surface of erosion between older rocks and overlying sedimentary beds is called an *unconformity*. Correct interpretation of unconformities is of great value in reconstructing the geologic history of regions.

Suppose that at some stage during the long period of slow erosion of a landscape under a declining energy gradient, tectonic uplift occurs and the land is elevated well above base level. Such a landscape is *rejuvenated*, and the former land surface becomes the initial form on which newly accelerated erosion operates. There is evidence that many landscapes have been so rejuvenated, some of them several times over. A rejuvenated landscape is more difficult to interpret than a landscape carved from a fresh, nearly featureless sea floor, because some of its forms may be *inherited* from the previous episodes of erosion.

Rejuvenation interrupts the former course of landscape evolution by injecting new energy into stream systems. The opposite interruption can also be visualized, that of a landscape that is lowered by tectonic movement closer to its ultimate base level of erosion. River systems would then lose energy.

Upward or downward movements relative to base level are called *interruptions* to the sequential evolution of a landscape because they change the rate of evolution, but not the ultimate forms. The orderly progression of landforms will be accelerated or slowed down, but not stopped.

If a landmass is uplifted vertically, rejuvenation begins near the former mouths of rivers and extends progressively inland until the entire river systems are graded to the new, relatively lower base level. If uplift is accompanied by tilting, an entire river system may be rejuvenated simultaneously by the increased down-valley slope at every point. Another river system, if it flowed opposite to the direction of tilting, might lose gradient and become a series of swamps and lakes by the same tilt. Many deductions have been made concerning the effects of various interruptions to landscape evolution, and many landscapes that show the effects of interrupted life histories have been described.

In addition to interruptions to their sequential evolution, landscapes also have *accidents.* These do not simply change the rate of orderly progression, but introduce entirely new processes and landforms. An accident permanently changes the course of landscape evolution.

The two types of accidents that befall landscapes are *volcanic* and *climatic.* The volcanic accident builds a new landscape on the old one, reconstructing slopes, damming rivers, and filling valleys. The volcanic accident is restricted to certain regions, especially around the rim of the Pacific Ocean and in a belt extending from the Philippines and Indonesia westward through the Mediterranean Sea. Volcanoes not only build great conical mountains like Fujiyama and broad domes like the island of Hawaii, but they also build lava plateaus that may cover more than 100,000 square miles. The Columbia Plateau of the northwestern United States, the Deccan Plateau of western India, and similar regions in central South America and southeastern Africa, all are built of numerous basalt lava flows totalling a mile or more in thickness.

Climatic accidents may be as dramatic as the advance of a glacier or as simple as a long drought. Previous chapters have emphasized the different degrees of effectiveness of weathering, mass-wasting, and erosional processes under various climates. If the climate grows cold, and snow accumulates as glacier ice, the erosion by glaciers makes a landscape very different from what may have been developing when water flowed in rivers. Aridity is another climatic accident that marks the landscape with distinctive landforms.

Many areas have been glacial or arid for a long time. In these regions, glaciers or deserts are not accidental, but are the normal result of the present planetary climatic zonation. Others areas that are humid today show strong evidence in soil, landforms, and sedimentary deposits that they were subjected to aridity or glaciation recently enough so that the marks remain. One of the most exciting areas of geomorphic research is the interpretation of climatic changes from landforms.

This section on initial landforms is easily summarized in Table 5–2. From the

Table 5–2

The Common Origins of Initial Landscapes

Tectonic uplift of former ocean floor	OROGENIC—mountain belts EPEIROGENIC—broad continental uplift
Interruptions to former development	REGIONAL UPLIFT: rejuvenation Without tilting—progressive rejuvenation With tilting—simultaneous rejuvenation REGIONAL SUBSIDENCE
Accidents	VOLCANIC CLIMATIC Glaciation Aridity

table it is obvious that there are many landscapes that can be taken as the initial form for a life history. Some start fresh and simple, and evolve only under the impact of the processes that can be observed at work today. Others, by far the most abundant, bear traces of former episodes of evolution that were disrupted or delayed. The skill and pleasure of a geomorphologist lie in reading as much of the history as he can.

Valleys: The Fundamental Units of Landscapes

We must constantly be reminded that landscapes evolve by loss of rock material. The unit of landscape is the valley, or the airspace from which the rock has been removed. Typically, valleys are the work of the rivers that flow along their floors; Playfair's Law states that principle clearly. If a valley is obviously too large or of the wrong shape for the river that now flows in it, we look for the explanation in the previous history of the region. Has the area been glaciated? If so, the present valleys may be the troughs carved by moving tongues

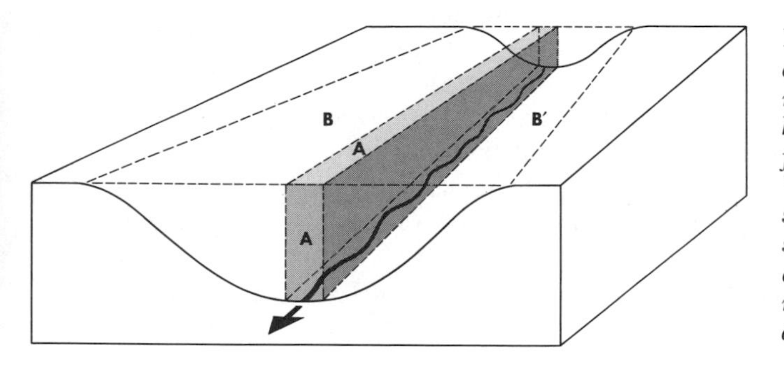

FIGURE 5–2 *A comparison of the volume of rock removed from a valley directly by stream erosion with that first moved by mass-wasting. Volume A was eroded by stream abrasion and corrosion; the larger volumes B and B′ were weathered and mass-wasted first, and the debris was later carried away by the river.*

of ice. What is the regional tectonic history? Perhaps the abnormal valley is a tectonic downwarp along which a river now flows. These are just two of the possible explanations for unusual valleys, but their very recognition as unusual emphasizes the rule that valleys are normally the work of the streams that flow in them.

Rivers do not erode the entire volume of their valleys (Fig. 5–2). The hydraulic geometry of streams shows that most of the sediment transported by rivers is delivered to them by mass-wasting. Only a small, but vital, amount of load is added from the stream bed through abrasion by transported particles or by chemical corrosion. This portion of the load is essential to landscape development because it is the downcutting that provides the slopes for mass-wasting.

Some rivers flow at the bottom of great canyons a mile or more deep, with sheer cliffs for valley walls. Such a canyon is primarily the work of a river that for some reason of energy supply or easily eroded rock has cut downward faster than mass-wasting could broaden its lateral slopes. One of the factors contributing to the Vaiont Reservoir disaster (Fig. 5–3; see also Chapter 3, Fig. 3–6) was the steep inner gorge, more than 1,000 feet deep, that the Vaiont River has cut since it was rejuvenated by tectonic uplift and increased discharge from Alpine snowfields. The Vaiont inner gorge has been cut in less than 18,000 years since the last glacier eroded the more broadly rounded outer valley, the floor of which was above the reservoir level. Rapid erosion by the river removed lateral support from the gorge wall and permitted sheeting joints to develop parallel to the wall by release of lateral pressure. The Vaiont Reservoir disaster is an example of mass-wasting keeping pace with valley deepening by a rejuvenated river.

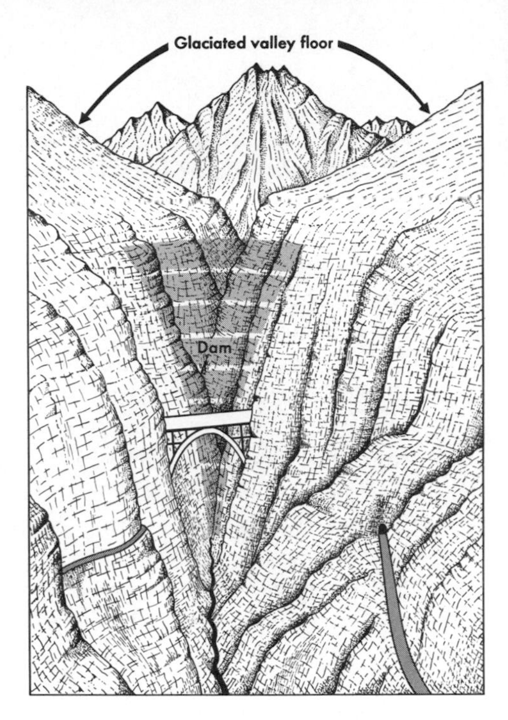

FIGURE 5–3 *The inner gorge of the Vaiont River, Italy, eroded in the floor of a more broadly rounded glaciated valley in the 18,000 years since the last glacier retreated. The dam is 875 feet high. (Redrawn from Kiersch, 1964.)*

Consequent, Subsequent, Antecedent, and Superposed Streams and Valleys

During the deduced early, or youthful, stage of landscape development, rivers rapidly expand their valley networks over the initial landscape. If the landscape were nearly featureless (Fig. 5–1), rivers initially follow advantageous hollows. Tributaries are few and much of the landscape may be poorly drained if initial slopes are gentle. Lakes may form in undrained depressions.

If orogenic deformation raises land above sea level, the ridges and troughs of deformed rock may control the initial drainage lines. If so, the drainage pattern will reflect the pattern of tectonic deformation. Rivers flowing either along chance hollows in a flat new landscape or between tectonic ridges are called *consequent rivers,* because their route is a consequence of the predetermined slopes. The valleys that they erode intensify the original relief of the surface. Tectonic troughs are eroded even deeper; tectonic ridges stand even higher above the eroded valleys.

As rivers erode deeper into a landscape they expand their valley networks headward. Water-gathering slopes (Fig. 3–9) focus water along their axes and cause valleys to form there. Weathering and soil formation are concentrated along poorly drained slope segments and prepare the way for gullies to be eroded. As the drainage net expands and organizes, it delivers water more efficiently to the master stream, which enlarges its valley still more.

If a landscape is underlain by layered sedimentary rocks, the erodibility of successive layers may be different. Under the same conditions of climate and initial slope, the tributary streams of an expanding drainage network that encounter easily eroded rock will cut deeper and longer valleys than those that encounter resistant rocks. *Structural control* then becomes a factor in the pattern of the developing valley system. Valleys that follow belts of easily eroded rock are called *subsequent valleys.* They may be formed by the headward erosion of streams along the exposure of easily eroded rock or they may be inherited from the pre-existing landscape. As the valleys of a new drainage network expand over a landscape, the initial consequent streams and their valleys may be replaced in importance by the rapid expansion of subsequent streams in structurally controlled valleys (Fig. 5–4). Usually the trunk stream retains its direction of flow consequent on the original landscape slope, even though it later may encounter resistant rocks that form local or temporary base levels for upstream segments.

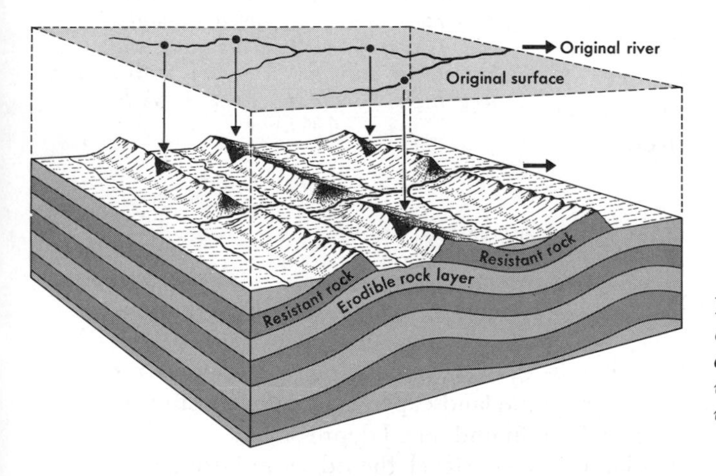

FIGURE 5–4 *Expansion of a consequent valley system over an initial landscape, with a later stage of erosion when subsequent tributaries have become dominant.*

An extensive descriptive terminology has developed for the patterns that rivers and their valleys make across landscapes. The most important kinds of drainage patterns are *deranged, dendritic, trellis, rectangular,* and *radial* (Fig. 5–5). The terms are purely descriptive, and apply to the map patterns of streams. *Deranged drainage patterns* are typical of newly emerged or newly deglaciated landscapes with gentle, irregular regional slope. Streams wander in disorder through lakes and swamps. *Dendritic patterns* are probably the most common for drainage networks. They simulate the "random walks" of moving water particles (Chapter 4) and indicate either lack of structural control or the presence of rocks of uniform erodibility. *Trellis patterns* are typical of subsequent tributaries eroded in belts of tightly folded sedimentary rocks. *Rectangular drainage patterns* often reflect either a regional pattern of intersecting joint systems, or a pronounced set of joints that cross belts of bedrock at a high angle. *Radial drainage patterns* are typical of volcanic cones or other sharp, high mountains.

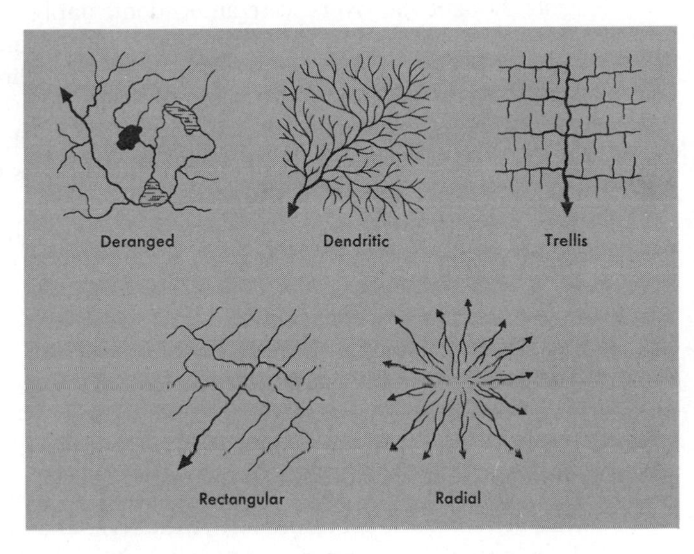

FIGURE 5–5 *Common types of drainage patterns.*

In addition to consequent and subsequent rivers and their valleys, two other genetic classes of rivers can be defined. Identifying them is of great value in working out the geologic history of a region.

In orogenic regions, the valleys that first form over a landscape may later have new tectonic ridges rise across them. The rivers may be defeated (forced to abandon their valleys) or they may be able to maintain gorges across the land mass rising in their paths. Streams that have maintained their valleys across tectonic ridges are said to be *antecedent*, for the river is older than the deformation. Antecedent rivers and their valleys are common only in actively orogenic regions. They form a third class of rivers and valleys, equal in importance to consequent and subsequent ones.

Rivers of the fourth class are called *superposed*. The word is a contraction of "superimposed," and implies that a river system has been let down or laid over a landscape. Suppose a land newly emerged from the sea has a thin cover of sediments that bury an older terrane of complexly folded rocks. On emergence, a consequent drainage system, perhaps with a dendritic pattern, might form. As the valleys deepen, they are eventually eroded into the old land under the covering beds. The dendritic pattern becomes *superposed* on the older rocks without regard to structural control. When the covering beds have been entirely removed by erosion, the only clue to superposition is the anomalous positions of the river valleys.

A common problem in geomorphology is to distinguish between antecedent and superposed rivers. In many parts of the Rocky Mountains, for instance, major rivers cross mountain ranges in narrow gorges, although a few miles away easy alternate routes are available around the ends of the ranges. Early geologists thought the rivers were antecedent, implying that the mountain blocks had risen very recently, after the rivers had become established. Later work has established that an interval of deep alluvial filling buried large parts of the Rocky Mountains during one stage of their formation. Rivers like the Yellowstone, Bighorn, and Laramie formerly flowed over the alluvium, and were superposed on the resistant rocks of the buried ranges as the area was regionally uplifted and dissected. The evidence for superposition is in remnants of the former alluvium now on the mountain sides at suitable altitudes, and the similarity of the alluvium on opposite sides of the ranges, which suggests a former continuous cover.

Sequential Evolution of Valleys

During the early, or youthful, phase of landscape development, when new drainage networks are rapidly expanding over the initial landscape, valleys are typically steep sided and V-shaped in cross section. Mass-wasting from valley walls delivers rock debris directly to the river channels. No orderly expenditure of the potential energy of the stream is possible, so long as sediment loads of irregular mass and grain size may be supplied at any point in the drainage system at any irregular interval. Only when the master stream of a region develops a continuous flood plain, and is thereby freed from unpredictable slumps or earthflows directly into the channel from adjacent valley walls, can the stream achieve the graded condition. Thus, the achievement of grade by the master stream of a drainage network, through its development of a continuous flood plain, is a signal event in the life history of a landscape.

By analogy with organic growth, where certain physiological changes mark maturity, a valley is said to be *mature* when the stream that flows in it is at grade. Typically, the graded river meanders freely across the alluvial flood plain of a mature valley, only infrequently undercutting the valley walls. Almost all of the sediment load in a graded river has been previously processed and transported by upstream tributaries. A mature valley is thus one that has an alluvial flood plain at least as wide as the *meander belt* of the river (Fig. 4–3). An observer on a hilltop can usually see upstream or downstream several miles along a mature valley, whereas a youthful valley winds sharply between the bases of interlocking noses or spurs extending from opposite valley walls.

Such waterfalls or rapids that were present in the early stage of valley development will have been removed by the time a valley reaches maturity. This deduction follows from the definitive requirement of a continuous flood plain. The downstream slope of the graded river will not necessarily decrease uniformly, however. Resistant rocks crossed by the stream will always produce a steeper channel slope in order to maintain continuity with the channel in more easily eroded materials downstream. Also, a tributary that enters a graded river may carry a greater proportion of load for its discharge than the master stream, and the master stream may maintain a steepened slope downstream from such a tributary in order to move the additional load. For that reason, it is said, the Missouri River has a steeper gradient below the junction of the Platte River in Nebraska. These variations in down-valley slope do not violate the definition of a graded river (and therefore a mature valley) as long as they are part of the interrelated and adjustable variables of the river's hydraulic geometry.

Sequential Evolution of Regional Landscapes

Ideally, the achievement of grade by the master streams that drain a region is one of two major criteria for maturity in the life history of the regional landscape. The other is complete integration of the initial landscape into the new drainage system. All slopes feed water into the first-order tributaries of the new master streams. No "unconsumed" or poorly drained residual upland surfaces remain from the initial landscape. The *local relief,* or altitude difference between adjacent high and low points, is at a maximum at this time. Divides between adjacent streams tend to be sharp-crested and straight-sided. Maturity is the most rugged stage of landscape evolution.

One of the complexities of defining maturity in the life history of a regional landscape is that the two criteria of maturity may not necessarily be met simultaneously (Fig. 5–6). If the tectonic uplift that created the initial landscape was very modest, the master streams soon achieve grade and flow in mature valleys while the tributaries are still expanding into undissected uplands (Fig. 5–6A). This condition of mature master valleys in a generally youthful landscape is typical of the Atlantic coastal plain of the United States south from New Jersey. The total potential energy of the rivers is not sufficient to maintain vigorous headward expansion of first-order tributaries, for the entire region is only a few hundred feet above sea level.

Conversely, if the initiating tectonic uplift of a landscape is excessive, the expanding drainage net will rapidly complete the mature dissection of the landscape while master streams are still downcutting in steep V-shaped youthful valleys (Fig. 5–6B). Most mountainous regions are in this condition. Complete dissection of the initial surface and maximum local relief, indicating regional maturity, coincide with steep, gorge-like youthful valleys.

Because of the possibility or even probability that regional and local stages of landscape evolution may not coincide, we usually evaluate the stage of sequential evolution of individual valleys independent of the stage of evolution of the regional landscape. One should always specify, when using the comparative terms *youth* and *maturity*, whether the regional landscape or a single valley is being considered. Because of the ambiguity that is possible in nonprecise usage, many geomorphologists dislike the terms youth and maturity, and prefer

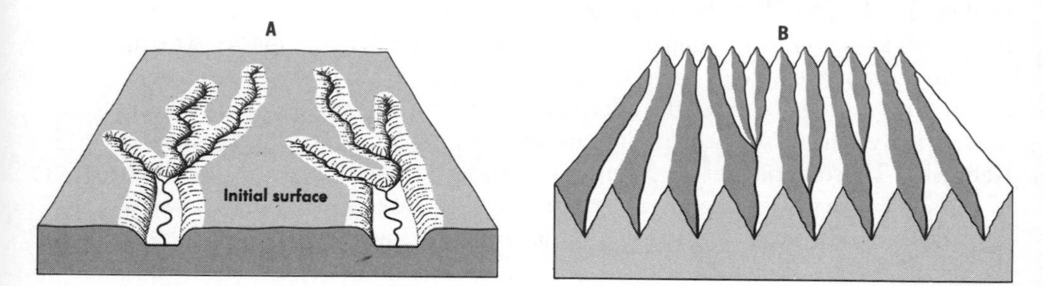

FIGURE 5–6 *Complexities in defining regional maturity of a landscape by the twin criteria of: (1) graded master streams in mature valleys, and (2) complete dissection of the initial surface. (Left) Slight rejuvenation. Mature master valleys in a nearly undissected plain. (Right) Excessive rejuvenation. Steep, youthful valleys in a maturely dissected mountain range.*

to describe the sequential evolution of both valleys and regional landscapes without analogy to organic growth. Nevertheless, to describe the Catskill Mountains of New York State as "a maturely dissected, glaciated plateau" conveys the maximum amount of information in the fewest possible words.

Old Landscapes and the Peneplain

If youthful and mature stages of valley and regional landscape evolution can be deduced, it is logical to suppose that an old or senile stage can also be deduced. Old age is more difficult and arbitrary to define than maturity, both in organisms and in landscapes. In living things, maturity is defined by achievement of reproductive ability, an arbitrary but useful definition. But when does an organism become old? Some loss of functions, some decrease in metabolic rate, or simply an intangible decline of ability are implied by the term "old," but the word is difficult to define precisely.

In landscapes, maturity is also well defined by the achievement of grade by the master streams, or the complete dissection of the former land, or both. But what happens next? Rivers at grade tend to remain at grade. Drainage nets that have covered the landscape can expand further only by chance intersection and *capture* of tributaries from adjacent networks; by the time mature dissection is achieved, watersheds of adjacent river systems are well defined and stable. Slopes and soils have become adjusted to climatic and vegetational conditions. But the rivers that drain such maturely dissected regions continue to carry sediment, so we know that erosional lowering of the land is proceeding. As an organism does, a landscape continues to "age" after maturity, but the criteria are difficult to define.

In an attempt to describe the form of landscapes that have undergone long-continued weathering and erosion in humid climates, W. M. Davis in 1889 introduced the elegant word *peneplain.* He used as a root the word "plain," in the geographical context of a regional surface of very low relief, near sea level. Realizing that base level is the limit of subaerial erosion, which like a mathematical limit is approached but not achieved, he prefixed to the word "plain" the Latin derivation "pene," meaning almost. Thus, *peneplain* was introduced to the scientific literature as a surface of regional extent, low local relief, and low absolute altitude, produced by long-continued fluvial erosion. The prefix "pene" converts the abstract concept of a perfectly base-leveled erosional plain to the concrete reality of a land surface.

Thus defined, the peneplain is the end form of erosion in a humid climate. Precisely, it is the "almost" end form in a sequential series that approaches but can never reach base level. As rivers erode landscapes lower and lower, the flowing water loses the potential energy of altitude, and the rate of further erosion decreases exponentially. The lower the altitude, the slower the rate of change. Somewhere in the sequential evolution of a post-mature landscape, the stage of the peneplain is reached.

Yet how can we define a landscape that is "almost" gone? Apparently, W. M. Davis had in mind that a peneplain was far from a mathematical *plane,* or even a *peneplane,* as some have proposed to spell the word. He envisioned a regional landscape, that is, one that includes the drainage networks of several master streams that enter the sea, in which the total relief is no more than a few hundred feet. One story is that he described a peneplain as a surface over which a horse could draw a carriage at a trot in any direction. In our modern era of excessive horsepower, it is hard to envision such a gently sloping surface of erosion.

Several criteria have evolved to help define a landscape in old age. By this stage of sequential evolution, both local relief and maximum altitude of the landscape will be very low. Slopes are deduced to be gentle and graded, summits very broad and gently convex, and soil or regolith very deep. Flood plains form a substantial proportion of the total landscape, and are at least several times as wide as the meander belts. Poorly drained areas are widespread, but they are on the flood plains and not on the uplands as they were in youth. Not only the trunk streams, but most tributaries are at grade. Structural control of drainage patterns become less noticeable in old age, for the differing erodibility of various rock types becomes insignificant when the potential energy of all streams is so low. The number of tributary streams decreases as broad, gentle slopes replace the numerous gullies of the youthful expanding drainage networks. Chemical weathering becomes progressively more dominant as slopes decrease and more rain water penetrates the ground.

On such old landscapes, hills will remain only on the divides between adjacent drainage systems, or on exceptionally resistant rock types. Commonly, both causes will contribute to the location of hills. A hill or low mountain that is a remnant of long-continued erosion was called by Davis a *monadnock*, from Mount Monadnock in southern New Hampshire, which was supposed to be of that origin. We now recognize that New England, including the region around Mount Monadnock, has had a much more complex history than simply the uplift of an old peneplain.

The peneplain is a necessary consequence of long-continued erosion by rivers, with accompanying weathering and mass-wasting. On theoretical grounds, the concept is valid if enough time is available. We have seen from measurements of erosional lowering of continents by rivers that the time is available, over and over again, in the decipherable part of geologic history. Furthermore, the geologic record of sedimentary rocks is full of unconformities that represent long periods of emergence and erosion of continent-sized regions. These unconformities are commonly nearly planar, although marine erosion during their final submergence probably swept away some of the relief that may have been present on the subaerial landscape just prior to submergence. The record of unconformities supports the theoretical validity of the peneplain as the end form of subaerial erosion.

It would be appropriate to describe a modern peneplain as a conclusion of this section, but unfortunately, none are known. The active tectonic movements and the repeated climatic fluctuations of the last million years or more have so complicated landscape development that no regional landscape yet studied is known to have reached old age without accident or interruption. Glacially controlled changes of sea level have been particularly influential in interrupting valley evolution in recent geologic time. The repeated fluctuations of sea level of several hundred feet have alternately rejuvenated and drowned the downstream ends of all rivers that enter the sea. The very parts of rivers that become graded first have been thus most affected by changes of sea level. Regions near the sea should be the places where peneplains begin to form, but every river that enters the sea today either has its lower part drowned as an *estuary*, or flows over an abnormally thick alluvial fill or delta that has accumulated because of the most recent rise of sea level and corresponding loss of stream gradient. Topographic maps of the lower Mississippi Valley are most often used to illustrate the features of the advanced stage of sequential landscape evolution, but they portray a surface of accumulation, not of stream erosion. The Mississippi River near its mouth flows on top of postglacial alluvium 600 feet deep, which has accumulated during rising sea level and tectonic subsidence. The only landscape that might qualify as a peneplain is the basin of the Amazon and Orinoco Rivers in South America. Unfortunately, the size and climate of the region have thus far prevented its proper exploration and description.

Theme and Variations

This entire chapter on deductive geomorphology has emphasized the life history or sequential evolution of landscapes that develop under conditions of excess rainfall and net runoff. The landscapes of the mid-latitude humid regions of the United States and western Europe have been so extensively studied that their processes of sequential evolution have been called the *normal processes* of erosion. The usage is poor, and implies that sequential evolution of a landscape under conditions different from those in Boston, Massachusetts, or Paris, France, is somehow "abnormal." It would be too time-consuming here to deduce the various stages of landscape evolution under conditions other than the "normal," or humid climate, but it is possible to suggest the sequential evolution of landscapes under various climatic conditions by briefly describing the deduced end forms. Most of the world remains to be explored by geomorphologists. The sketches and comments that follow only hint at the kinds of landscapes that are to be found in the lesser known and underdeveloped lands of the Earth.

For comparative purposes we will accept the peneplain as the end form of erosion in a humid climate. We can envision a regional profile of a peneplain as in Fig. 5–7A, with the vertical scale greatly exaggerated. Convex, low hills rise a few hundred feet above broadly concave lower slopes and flat flood plains.

The analogous end form in a semiarid region would be an assemblage of pediments (Fig. 5–7B). On approximately the same scale as the profile of a peneplain, an assemblage of pediments (*pediplain* of some authors) would have greater total relief and steeper slopes than a humid peneplain at the same stage of development. Wide pediments would sweep inland from arid coasts or river channels to head at knobs of resistant rock. This deduced end form requires through-flowing, if intermittent, streams to export the rock waste from the landscape. Some geomorphologists believe that the "pediplain" is the most common end form of erosion today. Whether they are correct or not remains to be proved, but the past importance of pediment landscapes cannot be doubted. So far as we know, grasses evolved in the Miocene Epoch, only about 25 million years ago. Without the remarkable fibrous turf of grasses, it is doubtful that the gentle summit convexities and slow soil creep, which now charac-

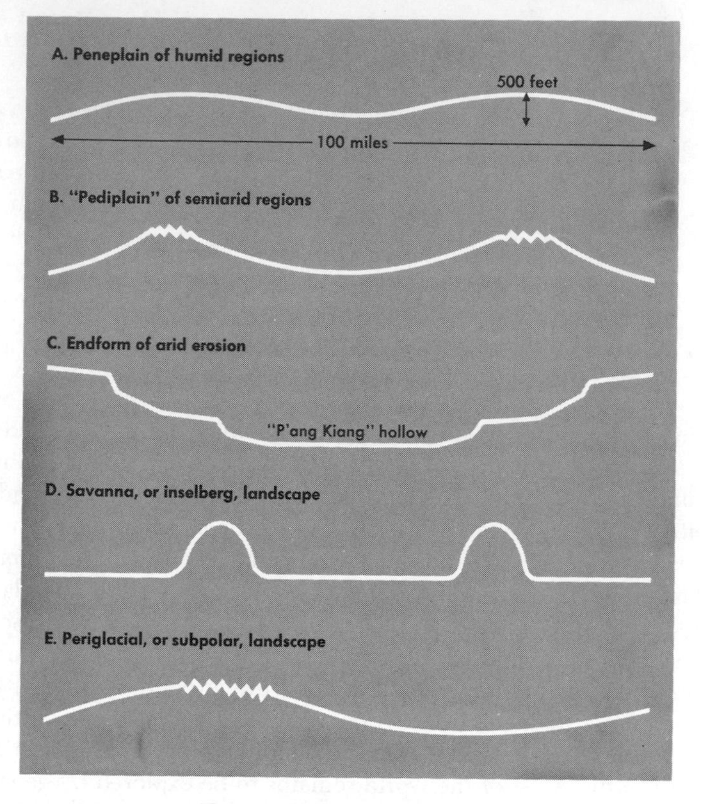

FIGURE 5–7 *The peneplain and climatic variants (hypothetical).*

terize humid landscapes, could be maintained. Prior to the Miocene Epoch, and the evolution of grasses, the flash floods and sheetwash that today characterize arid and semiarid regions may have been common even in nonforested humid regions. As one author has suggested, the deserts of the past may have been biologic, as well as climatic.

A third end form of erosion can be deduced for truly arid regions, from which no water flows to the sea. In profile (Fig. 5–7C) such a landscape might consist of *deflation hollows,* from which wind removes dust and sand, surrounded by cliffs, taluses, and alluvial fans that form by weathering and infrequent flash floods. Such a landscape in the interior of the Gobi Desert of Mongolia was described by C. P. Berkey and F. K. Morris in 1927, in their report of the remarkably successful Mongolian expedition of the American Museum of Natural History. The great "P'ang Kiang" hollows of the Mongolian desert are up to 5 miles in diameter and 200 to 400 feet in depth. Although they are surrounded by mass-wasting slopes and alluvial fans, little alluvium covers the central floor of the hollows, so it must be carried away by wind. No ultimate base level controls this landscape, in contrast to all other deduced end forms of erosion. The flat floors of the hollows are controlled by the ground-water level, which lowers by evaporation to permit the excavation of successively deeper hollows. Sea level is irrelevant to the evolution of a desert landscape.

A fourth end form of erosion can be deduced for tropical regions, especially the tropical savanna or seasonally wet-and-dry climate. In profile, the savanna analog of the peneplain is probably a series of very flat *wash plains,* above which "sugarloaf" hills, or *inselbergs,* rise sharply (Fig. 5–7D). During the rainy season, hundreds of square miles of these landscapes are under water, as surface runoff moves to the sea in wide, braided river channels that cover most of the landscape. In the dry half of the year, the savanna landscape is a monotonous sheet of alluvium dotted with thorn trees and sparse grass. The inselbergs, or *island hills,* stand like islands in a sea, and are the equivalent of the residual hills on peneplains.

The peculiar shape of inselbergs suggests a unique origin. During colonial times, German and French geomorphologists working in the savanna regions of east and west Africa, both north and south of the equator, concluded that the savanna landscape results from the "uncoupling" of weathering from erosion by the strongly seasonal climate. Chemical weathering penetrates to depths of several hundred feet, especially in granitic terranes, due to the excessive rainfall of the rainy season and high year-around temperature. Sheeting joints and other structural controls localize the deep weathering in such a way that large masses of unweathered rock may be bypassed by the advancing front of weathering. Erosion of the weathered debris is slower than weathering, so the land surface of broad wash plains cannot keep pace with the weathering front. As the wash plains slowly erode toward base level, unweathered masses of rock are exhumed, or excavated, and rise as inselbergs above the plains. On exposure, pressure-release and other weathering processes cause slabs or exfoliation sheets to fall from the inselbergs, which maintain their steep-sided form. As the many new nations of the savanna regions develop technical and academic personnel, the savanna landscape is sure to receive more of the study it deserves.

A fifth end form of erosion may possibly be deduced for regions of permanently frozen ground. Solifluction during the brief summer melt is the dominant form of downslope movement. Rivers flow so briefly that they cannot remove the mass-wasted debris from their channels. The deduced profile of an old subpolar landscape (Fig. 5–7E) might be characterized by very gentle, convex and concave slopes covered by thick solifluction debris passing at depth into unweathered, ice-saturated rock. Tectonic or other initial valleys are filled with alluvium and mass-wasted debris. Ridges are strewn with frost-wedged rock fragments.

Each of the five hypothetical end forms of subaerial erosion described above may be typical of at least 10 to 15 per cent of the land surface of the Earth. We know almost nothing about the life history of any of these landscapes except the peneplain, and little enough about that one. Add the observation that climatic change has repeatedly superimposed the conditions tending toward one end form on regions that were already evolving sequentially toward another end form, and the reasons for complexity in landforms begin to be apparent. Further add the complicating factors of different rock types, tectonic history, and the interference of man with natural landscape development, and the diversity of landforms becomes so probable that it should be a source of intellectual satisfaction that they can be grouped in any categories at all, much less a sequential series.

6

The edges of the land

The zone where land, sea, and atmosphere interact is almost a line, narrow in width and height but great in length. We call the zone of interaction the *coast.* Included within the coastal zone are both a narrow strip of land in which the proximity to the sea is felt, and the nearshore part of the sea in which the proximity to land affects the environment. The *shoreline,* or simply *shore,* is the more precise line of demarkation between land and water at any particular time. Lakes and ponds as well as oceans have shores, but the term "coast" is not applied to any region that does not border on the world-encircling ocean. Most coastal processes can be seen in operation on a smaller scale along the shores of lakes and can be studied there with profit, but in the broad scope of landscape evolution, lakes are doomed to be drained or filled, so the emphasis in this chapter will be on marine shorelines and coastal processes.

To this point of the book we have considered a landscape as an irregular surface that projects above sea level. At the coast, the land ends. The potential energy of falling water has become zero. Rivers drop their suspended and bed loads.

The river waters with their dissolved loads diffuse as sheets over the more dense ocean water, then mix and lose their identity in the great oceanic reservoir.*

At the coast, a new group of eroding, transporting, and depositing agents assumes the duties that rivers primarily perform "upcountry." Waves expend their kinetic energy against the edge of the land. Currents, generated by winds, waves, and tides, move sediment laterally along coasts, shoreward, or out into deeper water.

In the coastal environment, the tendency toward equilibrium between processes of change and the resulting landforms can once again be demonstrated. A major difference between the sequential evolution of coasts and of subaerial landscapes is that coastal processes operate only in the narrow vertical range from slightly above to slightly below sea level, along the edges of the land, whereas subaerial weathering and erosion operate over the entire surface of the land. Any change in level of either land or sea that is greater than the tidal range will interrupt the evolution of a coast and delay the attainment of an equilibrium coastal landscape. Coasts are much more sensitive to slight changes in level than are upland landscapes, and coasts are therefore likely to show forms inherited from previous episodes of development.

Energy Exchange at a Coast

Waves do most of the work of shaping coastal landscapes. Although romantic writers may imply that "the slow, relentless work of the tides" is the major process of coastal evolution, tides are clearly secondary to waves as agents of coastal modification. The principal geomorphic role of the tides is to change the water level so that wave energy can be expended over a greater vertical range. Secondarily, tides generate currents that erode, transport, and deposit sediment along coasts.

Sea and Swell Waves

Waves attacking the edges of the land represent a form of solar energy conversion that was not emphasized in Chapter 1. The differential heating of the atmosphere and sea by solar radiation produces atmospheric currents, or winds, that transport thermal energy poleward from the tropics. When wind blows over water, it produces surface waves in the water, which move in the direction the wind blows.

The mechanism whereby wind generates waves is not clearly understood. It is easy to visualize how a shearing stress by wind over water produces mass

* Another book in this series, *Oceans*, by Karl K. Turekian, discusses ocean water and the chemistry and geology of ocean basins.

transport in the surface water layer, but why the water surface should assume a wave form is not simple to explain. One theory emphasizes the inherent gustiness of wind as a cause of wave generation. Another theory explains wave propagation and growth as a result of slight pressure differences on the windward and leeward sides of small ripples as the wind blows over them. It is curious that such an obvious phenomenon as waves being kicked up by a wind is in fact such an elusive and complex process.

Wave generation represents direct transfer of kinetic energy from the atmosphere to the ocean surface. The energy chain from the Sun to a rocky headland being torn apart by waves during a storm can be diagrammed as follows:

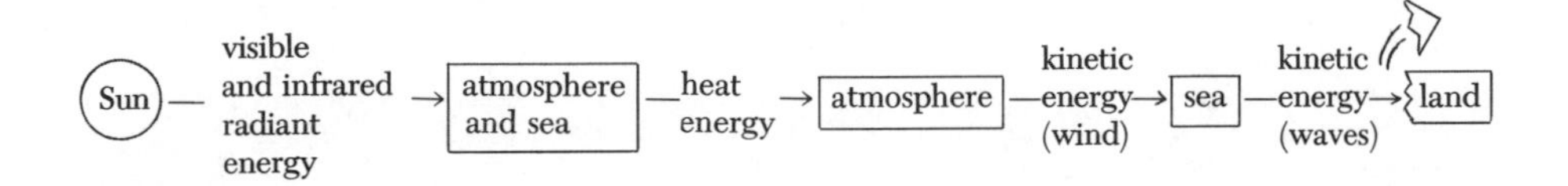

The geomorphology machine (Fig. 1–1) represents the combined work of waves and currents as a small grinder mounted close to the water's edge. The amount of solar energy expended by water waves is impressively large, but the zone in which it is expended is very narrow, and the total work done on landscapes by waves is small compared to that done by flowing river water.

In an area of strong winds and storminess over the ocean, the water surface is thrown into a confused mass of waves that intersect in peaks and troughs. A broad spectrum of wavelengths is generated by storm winds. Smaller waves and ripples run up the backs of larger waves. A stormy area of ocean surface is called a *sea* in mariners' terminology. When "a sea is running" it is best for ships to avoid the area. Waves as high as 75 feet have been reported, but accurate observations of such waves is understandably difficult.

Radiating outward from the generating area of a sea, waves become more orderly. Surface waves with the longest wavelengths advance most rapidly, and successive wave trains sort themselves out as they move away from the generating area. The longest ocean waves are generated in the large and stormy oceanic areas between latitude 40° S and Antarctica. These long waves may travel thousands of miles before they strike a coast. They have been identified as the cause of erosion in both California and the British Isles.

The regular pattern of smooth, rounded waves that characterizes the surface of the ocean during fair weather is called *swell.* Swell is usually composed of several wave trains of different wave lengths, often moving outward from more than one generating area. From the air, swell looks like a grid of intersecting lines. The individual waves in obliquely intersecting wave trains may alternately reinforce and cancel one another. If the waves in two wave trains are traveling in nearly the same direction at nearly the same velocity, the period of their mutual interference may be quite long. It is a common saying that every seventh wave that reaches the shore is larger than the others. That is not true, but it is true that swell approaches a shore as a complex of wave trains that combine to produce alternately a succession of higher waves and a succession of lower ones.

Swell waves transmit energy outward from a stormy area in two forms, potential and kinetic. The height of a wave determines the potential energy of position above still-water level. The motion of individual water particles as a wave passes is a measure of the kinetic energy of the wave. In deep water, waves move forward continuously, but a marker on the water surface would be observed to rise and fall in a circular path with only a slight net forward motion as each wave passes (Fig. 6–1). The potential energy of the wave moves forward with the wave form, but the kinetic energy of each moving water

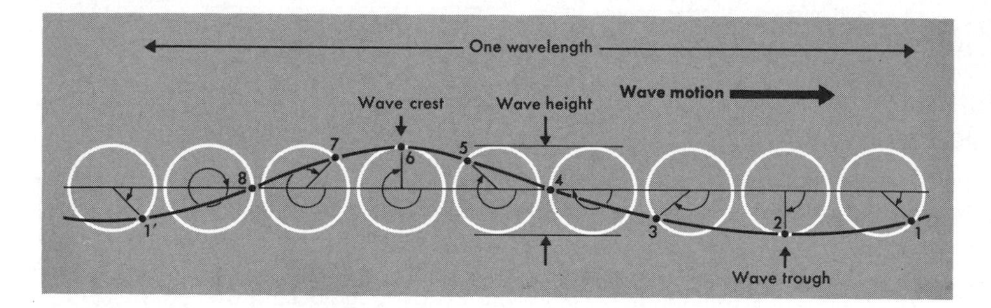

FIGURE 6–1 *Motion of a wave contrasted to the motion of eight numbered surface water particles as the wave passes. In this construction, each water particle moves in a closed, circular orbit as the wave passes. The net energy transmitted forward by the wave would be measured only by the wave height, for no mass of water is displaced forward permanently. In reality, the orbit of each water particle is not quite circular, but has a slight forward component in each rotation.*

particle is expended in the nearly circular orbit of the particle. Thus, the total energy of a wave train moves forward less rapidly than the speed of the waves. Each wave appears to die out at the front of a wave train as a new wave forms from it to advance the front.

The orbital diameter of water particles in a wave decreases rapidly with depth, in a geometric progression related to the wave length. The orbital diameter is halved for each increase in depth equal to one-ninth of the wavelength. Thus, at a depth equal to the wavelength, the orbital diameter is about 1/512 of the surface diameter, which is the wave height. A swell wave 100 feet long and 4 feet high would stir a water particle at a depth of 100 feet through a small vertical circle less than 1/10 inch high. This rapid decrease in water motion with depth explains why a submarine can ride out the most severe storm by submerging 100 feet or more, for most surface waves have wavelengths less than a few hundred feet. We regard a wave as moving in deep water when the water depth exceeds one-half the wavelength. At that depth, the drag of the wave on the bottom is negligible.

Surf and Breakers

As deep-water swell approaches a coast, the form of the waves and the method of water and energy transport change. As the water depth decreases, the orbital motion beneath the waves is distorted from a circle to an ellipse, then to a to-and-fro linear motion. Sediment on the ocean floor is moved back and forth by the waves, and absorbs energy from the moving water. At this depth, the conversion of wave energy to geomorphic work begins. Although very long surface waves may stir water and bottom sediment to great depths, ordinary waves do not directly affect bottom sediment transport or erosion at a depth greater than about 30 feet. This depth, which has been called *wave base,* is a convenient reference level, although we should keep in mind that the work of waves does not abruptly cease at some certain depth.

At the surface of the sea over a shoaling bottom, the wave velocity is decreased by bottom friction. The wavelength is correspondingly shortened as deep-water waves farther seaward continue to move forward at full speed. The wave height also increases as the wavelength shortens. The orbits of individual water particles in the wave change from nearly circular to strongly flattened ellipses. As the crest of each wave approaches, the water moves rapidly forward as well as upward. Ultimately, the forward motion of the surface water mass equals the decreasing forward motion of the wave front. The wave assumes a steep face, then the crest of the wave collapses forward into the trough. The deep-water swell becomes *surf,* or lines of *breaking waves.*

Breakers form in the surf zone where the bottom depth is about one-third greater than the breaker height. Several lines of breakers may form, each successively lower train of waves breaking nearer the shore. On gently sloping coasts, the first line of breakers may be as much a mile offshore, and the incoming water may re-form into lower waves that break again at shallower depths.

Most of the wave energy is expended in the surf zone. Deep-water swell can be reflected by a vertical seawall or steep cliff with almost no transfer of energy from the waves to the structure, because the mass of water in each wave is not moving forward significantly. However, when a breaking wave hits a cliff or seawall, thousands of tons of water are in motion against the structure. The greatest pressures are exerted by breaking waves that curl over at their crest and trap air between the wave face and a steep wall or cliff, so that the air is compressed; shock pressures of 12,700 pounds per square foot have been recorded against seawalls. The duration of such great pressure is less than a hundredth of a second, but large blocks of rock can be pried loose and moved by the repeated assault of the waves.

The wave energy against the coral reef of Bikini Atoll was carefully calculated by W. H. Munk and M. C. Sargent in 1954. Assuming that the average breaker height against the exposed reef was seven feet, they concluded that the waves' total power (continuous rate of doing work) against the reef was 500,000 horsepower, and reached eight horsepower per linear foot of reef in the most exposed locations. For comparison, Hoover Dam has a hydroelectric output of about 1,800,000 horsepower.

Breaking waves not only exert great force against coasts, but the moving water also drags rock fragments rapidly over fresh rock faces and abrades both the particles and the bedrock. Bricks and broken glass are smoothed and rounded by only a few days of abrasive tumbling in the surf on a sandy beach.

Each train of waves in deep-water swell approaches the coast in parallel lines, equally spaced like ranks of well-disciplined troops. When some portion of a wave front begins to "feel bottom," however, the wave is *refracted*. Sup-

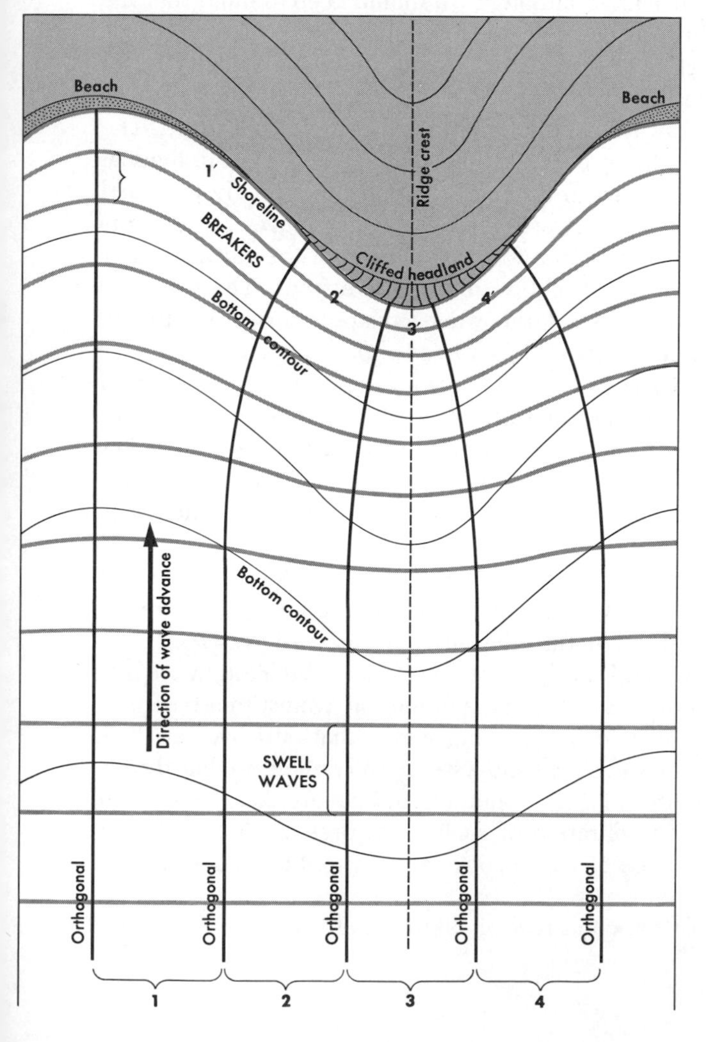

FIGURE 6–2 *Wave refraction over an irregular shoaling bottom. See text for a description of the figure.* Orthogonals *are lines drawn perpendicular to the wave crests, equally spaced along the swell so that segments 1 to 4 have equal amounts of energy.*

pose a portion of a coast consists of a sloping ridge or nose of land extending out under the sea (Fig. 6–2). A *headland* forms where the ridge protrudes into the sea, and a submarine ridge continues offshore into deep water. As the parallel lines of swell waves move over the submarine ridge, a segment of each wave in turn is slowed by bottom friction. The crest of the wave in deeper water on each side of the submarine ridge continues to move forward at its former speed, so the wave front becomes concave landward, and the wave energy converges toward the headland. Wave refraction over a shoaling submarine ridge thus focuses wave energy against the headland shore.

Conversely, where swell approaches the coast over a submarine hollow or valley (Fig. 6–2), the wave front continues to move forward at full speed over the deep part of the valley, but is dragged back on either side. The wave front becomes convex landward, the crest of the wave is stretched or attenuated, and the energy of the wave is diverted from the axis of the submarine valley.

If a wave is of uniform height along its crest, equal lengths of wave crest contain equal amounts of energy. Then if part of a wave front converges over a submerged ridge, as in Fig. 6–2, the crest length of that segment is shortened and the wave energy is concentrated. The wave increases in height as it nears the shore. When it breaks (Fig. 6–2, segment 3′) its energy is concentrated on a short length of shoreline and a relatively large amount of geomorphic work is done on the land.

Over the comparatively deep water in a cove or bay, the crest length of a deep-water wave segment increases, and the wave height correspondingly decreases. When the wave breaks on the shore of a cove (Fig. 6–2, segment 1′), it may be little more than a line of curling foam, although breakers many feet high may be pounding the headlands on each side of the cove within full view of an observer on the sheltered beach.

Wave refraction is the basis for two important generalizations about the evolution of coasts. First, initial seaward protrusions of the shoreline caused by submarine noses of sloping ridges tend to be eroded faster than adjacent coves formed by the shoreline at the head of submarine valleys. Wave refraction tends to simplify an irregular initial shoreline by removing the headlands.

The second effect of wave refraction is to generate currents flowing along the shore from headlands, where the focused breakers raise the water level, to the axes of adjacent coves, where the water level is lower. These *longshore currents* transport the sediment eroded from headlands into adjacent coves where beaches are built. The simplification of a shoreline by wave refraction thus involves both the filling in of indentations as well as the erosion of headlands.

In theory, a shoreline should tend toward a straight line parallel to the wave fronts of the dominant swell. In fact, headlands are commonly composed of a more resistant rock than the shores of adjacent coves, and even though wave attack is concentrated on the headlands, the more resistant rocks continue to form seaward bulges as the entire shoreline retreats. Straight shorelines are characteristic only of coasts formed by rocks of uniform erodibility.

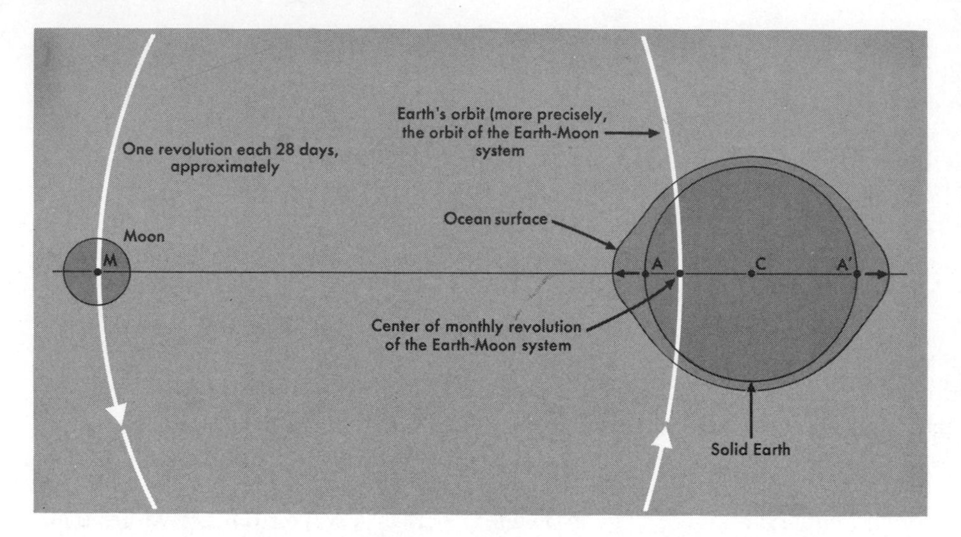

FIGURE 6–3 *The theoretical equilibrium lunar tide. The distance between the center of the Earth and the Moon is approximately 60 Earth radii. Because gravitational force is inversely proportional to the square of distance, the lunar gravitational force at A is greater than at C by the ratio* $(60/59)^2$*, and at A′ the force is less than at C by the ratio* $(60/61)^2$*. Therefore, with reference to the center of the Earth, lunar gravity exerts a 3.4% greater force* toward *the Moon at A, and a 3.3% greater force* away *from (or a lesser force* toward) *the Moon at A′. These forces, with the addition of a small centrifugal effect of the monthly revolution around the common Earth-Moon center of gravity, tend to raise a cigar-shaped bulge in the oceans, one apex centered beneath the Moon and the other on the opposite side of the Earth from the Moon.*

Tides

Tides are a unique form of energy input into the geomorphology machine. They are the result of the gravitational attractions between the Earth and the Moon and Sun. No other celestial objects are close enough to the Earth or large enough to create tides. As was noted in Chapter 1, the Moon is more than twice as effective as the Sun in raising tides on Earth.

If the Earth were entirely covered with deep water, an ideal or equilibrium tide a few feet high would be generated by the Moon and Sun. The major effect would be to raise two high tides on the world ocean, one centered beneath the Moon and the other on the side of the Earth farthest from the Moon (Fig. 6–3). As the Earth makes one full rotation eastward on its axis each 24 hours and 50 minutes with reference to the Moon, a high tide would appear to sweep westerly over the ocean every 12 hours and 25 minutes. Every 6 hours and 12 minutes after a high tide passed any point, a low tide would occur. The equilibrium lunar tide would be alternately reinforced and partially canceled by the solar tide, similar in form but less than half as high. The result would be a *semidiurnal* (twice-daily) tidal cycle, 20 per cent greater in vertical range when the solar and lunar effects were in phase during New Moon and Full Moon, and about 20 per cent less in vertical range when the Moon and Sun are in quadrature, or 90° apart in the sky.

The concept of an equilibrium tide is useful to show the forces that generate tides, but real tides in the world oceans are very different. For one reason, the semidiurnal equilibrium high tides would have to sweep westward through equatorial oceans at about 1,000 miles per hour in order to stay directly beneath and directly opposite the Moon. Water cannot be displaced over the surface of the Earth that fast. Secondly, the equilibrium tide would have the form of a standing wave with a wavelength equal to half the Earth's circumference. We have seen that the depth at which any wave "feels bottom" is proportional to the wavelength. The oceans would have to be over 14 miles deep to permit the equilibrium tide to move with a minimum drag on the bottom, yet on the average the ocean is less than one-fifth that deep. The result of the friction in water and on the ocean bottom is to prevent an ideal equilibrium tide from developing.

Real tides develop in oceans primarily as a response to the lunar and solar gravitational forces, but their periods and heights are affected by the size and depth of the various ocean basins, the shape of the shorelines, and the latitude of the basins, among many other variables. Each ocean, gulf, or inland sea has its own tidal pattern. Tidal predictions in almanacs and newspapers are based on mathematical analysis of the previous records at some seaport, rather than on any general theory of tidal motion.

Tidal ranges are high only where the tides enter a semi-enclosed sea or gulf that has the shape and depth to produce a natural period of oscillation of water that is in resonance with some period of the tide. The greatest tidal range on Earth occurs in the Bay of Fundy, Canada, and may vary 53 feet between high and low water. The Bay of Fundy has a natural period of oscillation of slightly over six hours, almost in phase with the ideal semidiurnal tide. The result is comparable to the disastrous effect of trying to carry a shallow tray full of water in your hands. A wave begins to oscillate back and forth, and if you happen to tilt the tray in the same direction that the oscillating wave is traveling, the water is sure to spill over the low end. In the Bay of Fundy, the high tide grows in height from about 10 feet to over 50 feet as it travels northeast into the head of the bay.

The high tidal ranges that result twice each month during the New Moon and Full Moon, when the lunar and solar tides are in phase, are called *spring tides*. When the lunar and solar effects partially cancel each other, during the First and Third Quarters of the lunar cycle, the tidal range is at a minimum. These tides are called *neap tides*. Other seasonal changes in the orientation of the Earth, Moon, and Sun in space amplify or diminish the range of tides. Seasonal climatic changes such as persistent onshore or offshore winds also greatly influence the level and range of the tides.

In addition to their primary geomorphic role of raising and lowering the level of wave attack against the land, tides also generate strong currents that may flow at speeds of five miles per hour or more. The most rapid currents develop in narrow channels that connect basins with different tidal periods. Hell Gate in the East River at New York City is one such place. The high tide in Long Island Sound takes almost three hours to move westward through the Sound. In that time the tide in the less restricted New York Harbor has crested and is half way down to low water again. Because of the difference in level, the water in the Sound ebbs south through Hell Gate at speeds of five knots (six statute miles per hour). About six hours later the tide in the west end of the Sound is low, but the water in New York Harbor is nearing high tide level and the current floods north into the Sound almost as fast as it previously ebbed. The name of the strait is an obvious reference to the violent and changeable tidal currents.

Occasionally where a tide enters a large river mouth, it begins to drag the bottom like any long wave and steepen its forward slope. In extreme examples, the river mouth develops a *tidal bore,* a single wave that races upstream at 10 to 20 miles per hour as a vertical wall of water as much as 10 feet high.

Tidal currents and bores are obviously capable of extensive erosion and transportation of sediment. Deep submarine channels regularly mark the paths of tidal currents through narrow straits. Under suitable conditions, tidal currents can scour the ocean floor and transport sediment at depths much greater than are normally affected by waves. Especially in semi-enclosed basins where waves are small, tides and their currents become major geomorphic agents.

Organisms

The energy environment of coastal waters would not be complete without a mention of biological activity. Where rivers enter the sea, they bring dissolved mineral matter and organic debris that causes a proliferation of life. The oceanic environment is thermally and chemically much more stable and suitable for lower forms of life than is river water, but organisms in the coastal sea or in an estuary have the best of both worlds. Animals filter great quantities of suspended matter from river water and deposit it as mud from their digestive tracts. Diatoms and other plants remove much of the dissolved silica and other compounds from river water as it begins to mix with the sea.

In coastal water, most of the weathered rock from the land begins a series of "unweathering" processes that eventually create new rocks from the debris of older ones. Many of the processes involve organisms. On many coasts, tidal marshes incorporate dissolved minerals from river water into plant tissue that may either be deposited as peat or eaten by other organisms as the first step in the complex food chains of the oceans. Coastal plants and animals build reefs, fill lagoons with peat and entrapped mud, and secrete protective crusts over loose sediment. They also gnaw, dissolve, or burrow coastal rocks and sediment, and thereby become significant agents of erosion. All of the energy of these plants and animals is derived ultimately from the Sun. One of the most amazing features of corals and related organisms is that they flourish best on the exposed windward faces of tropical reefs, where the intense agitation of breakers brings them the greatest supply of dissolved nutrients. Instead of being destroyed by the tremendous horsepower of the breakers, they use the power to manufacture tons of new limestone for their foundations.*

Coastal Sediment

If some wave energy was not absorbed in the work of moving sediment along coasts, waves would accomplish far more erosion. Given enough time and no burden of sediment, waves would cut wide, gently sloping submarine benches across the edges of the land, backed by rugged cliffs. As it happens, a great deal of the sediment derived from the erosion of sea cliffs and a much larger amount of sediment from the mouths of rivers remain within the shore zone for a long time, gradually being abraded to finer grain sizes and swept away, either seaward or back onto the land. This sediment in motion along a shore is the *beach.* A beach performs the same geomorphic function as the flood-plain alluvium in river valleys; it absorbs energy and therefore moves during times of storm, and thereby stabilizes the rate of energy conversion into geomorphic work. A continuous beach along a coast is as good a criterion of the graded condition as is a continuous flood plain on the floor of a river valley or a continuous sheet of sediment on a semi-arid pediment.

Since coastal sediment is provided by rivers, waves, wind, and organisms, we will first examine the sources of coastal sediment and the kinds of sediment provided from the various sources. Then we will consider the various processes of sediment distribution. Subsequently, some of the landforms that are built of coastal sediments will be described.

Sources of Coastal Sediment

Most of the sediment on coasts was transported there by rivers from the eroding landscapes. Direct wave erosion accounts for only perhaps 10 per cent of the material moving along coasts. Wind blows sand and dust into coastal waters downwind from desert regions, but the mass contributed by the wind

* The alteration of sediments by organisms is described in greater detail by Léo F. Laporte in another book in this series, *Ancient Environments.*

is very small. Waves bring sediment onshore from deeper water, but most of this sediment was previously carried to the sea by rivers.

The grain size and chemical composition of sediment brought to coasts by rivers are primarily determined by the nature of the bedrock over which the rivers have flowed. In addition to this primary control, sediment is further sorted by size and composition during the process of river transport. Large rivers, like the Nile, Mississippi, Ganges, or Colorado, reach the sea carrying primarily silt and clay-sized debris, or mud. Shorter, steeper streams may carry sand or even gravel to the coast.

Wave erosion of sea cliffs initially produces poorly sorted sediment. As the waves cut away the foot of a cliff, landslides bring down great masses of unsorted, broken rock to the shore. However, waves and nearshore currents are particularly efficient at sorting sediment by size and specific gravity. During a short walk along the shore away from the foot of an eroding sea cliff, you can see beach sand or gravel becoming finer in size, more rounded, and better sorted. Therefore, wave erosion and transport tend to accomplish in a short distance along a shore what a river may need hundreds of miles of flow to complete: the separation of sediment into various grain-size categories that move to different parts of the coastal zone depending on the available transporting agents and energy supply.

On some coasts, organisms provide most or all of the sediment supply. A coral reef may build up to the surface of the tropical ocean. Through erosion of the reef front, coral blocks, sand-size grains, and limy mud are made available to be built into islands on the reef surface. Sandy little islands each less than a mile long and no more than 10 to 15 feet above sea level commonly are the only dry land on a coral reef or atoll.

Along the Everglades coast of Florida, as on many other coasts that are not backed up by eroding highlands, broken shells and lime mud of marine microorganisms form the entire supply of sediment from which the land is built. In addition, the mangrove trees that grow in the brackish shallow water of the Everglades and many tropical coasts trap sediment washed in among their roots. To this entrapped debris is added the fallen leaves and branches of the mangroves. In only a few years mangroves and similar plants can build new coastal land out of accumulated peat and mud.

Sediment Transport

Because of the slight net forward movement of water in waves (Fig. 6–1) the shoreward velocity of water orbiting under the crest of advancing waves is greater than the seaward velocity under the trough of the waves. Thus, as waves feel bottom they tend to move loose sediment landward, unless the offshore slope is too steep. Depending on the nature of the bottom and the form of the waves, waves can move sand onto beaches or erode it away. Waves with long wavelength are particularly likely to bring sand from offshore toward the shoreline. Short, steep waves from local storms are more likely to stir up sediment from the nearshore region and keep it in suspension until it settles into deeper water.

The sand brought into shore by waves is commonly river sand that was carried seaward and dropped as the river water mixed with the sea. A complicating factor in identifying the sources of coastal sediment is that during continental glaciation, when sea level was lowered several hundred feet, rivers flowed across coastal plains that are now submerged. K. O. Emery estimated in 1968 that 70 per cent of the sediments on the continental shelves of the world were deposited in a different kind of environment. Much of the sediment which is now moving along coasts is *relict*, that is, it was deposited by rivers or glaciers above sea level, but is now submerged by the postglacial rise of sea level. A large part of the relict sediment that is now being moved along coasts was brought to the coastal regions during times of different climatic conditions. This sediment was not transported by the waves and currents that are now reacting with it. Hence, the grain size and composition of coastal sediments today commonly reflect past environments more than they reflect the present one.

Whether brought in from deeper water by waves, or out of the mouths of rivers, or eroded from headlands, sediment in the nearshore zone is moved along the coast by longshore currents generated by waves, winds, and tides. The collective term for the movement of sediment along coasts is *longshore drifting.* Most of the sediment is moved in the surf zone, and that part of the longshore motion is called *littoral drifting.* A smaller amount of sediment is moved by the oblique wash of waves on beaches when the waves are approaching the shore at an angle. This process is called *beach drifting.* It is readily observed by tossing a floating object into the water and watching the saw-tooth pattern it traces up the beach at an angle, bobbing on the incoming water, then stopping and washing or rolling back down the sloping beach perpendicular to the shoreline.

Littoral drifting is hard to observe, for most of the action takes place beneath the breaking waves in the surf zone. Recalling that breakers involve the forward rush of a large mass of water, you can recognize that surf actually "piles" water into the nearshore area, especially on headlands. When waves approach at an angle, the excess water mass flows away parallel to the shore as a strong *littoral current,* just inside the surf zone. Sediment thrown into suspension by a breaking wave is carried by the turbulent littoral current. Eventually, the littoral current turns seaward along the trough of a submerged valley, or breaks through the incoming surf as a *rip current.* Plumes of suspended sediment sweeping seaward through the surf zone mark the paths of rip currents.

Along many coasts there is a dominant direction of wave approach and regional currents, and therefore a dominant direction of longshore sediment drifting. On most of the southern California coast, the dominant movement is southward. On the south shore of Long Island, New York, the drifting is toward the west. In northern New Jersey, the dominant drifting is northward into lower New York Bay. On the remainder of the Atlantic coast of the United States, from northern New Jersey to the Florida Keys, the dominant longshore drifting is southward.

Transport of sediment in ocean currents, as in rivers, is determined by the specific gravity of the sediment grains and their size. Heavy minerals such as gold, magnetite, and garnet settle out of flowing water quickly, whereas lighter-weight minerals such as quartz and feldspar remain in suspension longer and are transported farther. More important, for sediment grains coarser than fine sand (average grain diameter, 0.2 mm) the current velocity necessary to erode and transport the grains increases as the square root of grain diameter. That is, when current velocity is doubled, particles four times as large are set in motion, and when current velocity is tripled, particles nine times as large are transported. No surprise, then, that on rocky headlands, where wave erosion and longshore currents are strong, only large boulders litter the base of cliffs. Everything finer has been swept into coves adjacent to the headland, or into some other sheltered place. Fine sand moves easiest and farthest in longshore currents and therefore becomes the most common material on beaches.

Sediment grains smaller than fine sand can be easily transported by weak currents once they are put into suspension. However, the exceptional property of very fine sand and mud is that if they are allowed to settle out of water, the particles cling together as a tough, slick mass that resists renewed erosion. A silt or clay bank may withstand erosion as well or better than a coarse gravel bank.

The cohesiveness of fine sediment helps explain why beaches are largely sand. When surf breaks, the turbulent energy released throws all sizes of sediment into suspension. Now, visualize a rising tide, with tidal currents flooding into estuaries and bays. Gravel and sand quickly settle out of the water in the surf zone or on a beach. Suspended mud stays in suspension longer and is carried into a protected part of a shallow tidal cove, where some of the mud slowly settles to the bottom during the interval of negligible currents as the tide turns. As the tide begins to fall, the tidal currents increase again, this time ebbing out of estuaries and bays. As the current velocity increases, sediment is eroded, but all of the mud that was deposited during a certain current velocity a few hours earlier is not picked up again by the same current velocity during ebb tide. The particles cling to the bottom until a current of considerably higher velocity sweeps over them. Each cycle results in a net gain of deposited mud.

The cohesiveness of mud and the finest sand results in the concentration of these grain sizes in the quietest parts of the coastal environment. Usually these quiet places are in tidal marshes, lagoons, estuaries, or bays. Along coasts where fine-grained sediment is abundant, great *mud flats* are exposed at low tide. If the intertidal mud flats are colonized by salt-tolerant grasses, tidal marshes result. The total effect on coasts is that tidal currents sweep mud into bays and estuaries and leave pure, clean sand in the surf zone of exposed beaches.

Coastal Landscapes

Coastal landforms have both erosional and depositional origins. Dependent as they are on the regional tectonic activity and on the nature of the bedrock, climate, river discharge, average wave height, tidal range, and many other variables, coastal landscapes provide the most beautiful and varied scenery in the world. Given enough time, coasts will evolve sequentially toward stable equilibrium forms that will be generally smooth and regular. Fortunately for those who enjoy coastal scenery, the geologically recent rise of sea level by several hundred feet in the last 15,000 to 20,000 years has initiated new coastal landscapes all over the world. Only coasts made of easily eroded materials such as sand and gravel, or weakly cemented sandstone, have been exposed to sufficient wave energy to develop well-adjusted forms. On most rocky coasts, the waves have barely begun their work. Perhaps it is this lack of adjustment and the resulting clashes of wave energy and rock that make coasts such interesting places to visit.

The basic topographic elements of steep coasts are sketched in Fig. 6–4.

FIGURE 6–4 *The basic topographic elements of a steep coast. The* wave-cut bench *and* wave-cut cliff *are erosional. The* wave-built terrace *is depositional. The upper surface of the wave-built terrace is the* shoreface, *on which most sediment movement takes place. The thin, wedge-shaped layer of sand that extends landward across the intertidal zone from the shoreface and is in active motion on the wave-cut bench is the* beach.

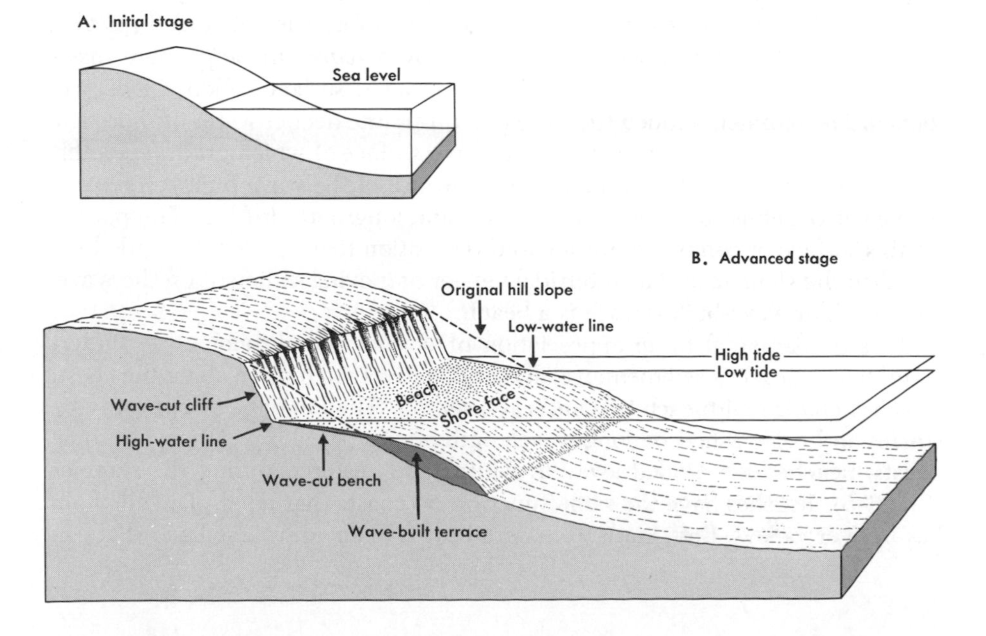

FIGURE 6–5 *Boomer Beach, near La Jolla, California. The beach sand accumulates annually during the summer (left), but is swept away each year by winter storm waves (right). (From* Geology Illustrated *by John S. Shelton. W. H. Freeman and Company. Copyright © 1966.)*

When water level comes to rest against sloping land through any change of level of either land or sea, a new sequence of landforms is initiated. Typically, the waves cut an intertidal notch, backed by a *wave-cut cliff* where mass-wasting is very active, and floored by an abraded surface called a *wave-cut bench.* The sediment eroded from the cliff drops into deeper water offshore and accumulates as a *wave-built terrace,* the upper surface of which grades smoothly shoreward into the wave-cut bench. The surface of the wave-built terrace and wave-cut bench is the zone of surf action and longshore drifting. The portion of this surface of combined erosion and deposition that is below low-tide level is called the *shore face.* The intertidal veneer of moving sediment on the wave-cut bench or wave-built terrace is a beach.

It is fundamental to an appreciation of coastal geomorphology to understand that a beach is sediment, usually sand, actively in motion along the coast. Beaches change with each tide and each season. They may be swept away by storms and replenished by long-wavelength waves of calm weather. On some coasts, beaches are eroded away in winter and restored during the summer (Fig. 6–5). Beaches have the same function on coasts that flood-plain alluvium has in river valleys. Both kinds of material provide an adjustable layer that can

quickly respond to short-term changes in energy input and thereby equilibrate the rate of development of long-term sequential landforms.

When the coast is a region of gently sloping land ("flat" will be used subsequently), the resulting landscape is unlike that of a steep coast (Fig. 6–6). Whether the region is a former flat seafloor that emerged or a former flat

FIGURE 6–6 *Typical landforms of a flat coast. Most of the features are depositional, for wave energy is largely expended in moving loose sediment as waves pass across the gentle offshore profile, and little erosion can be accomplished.*

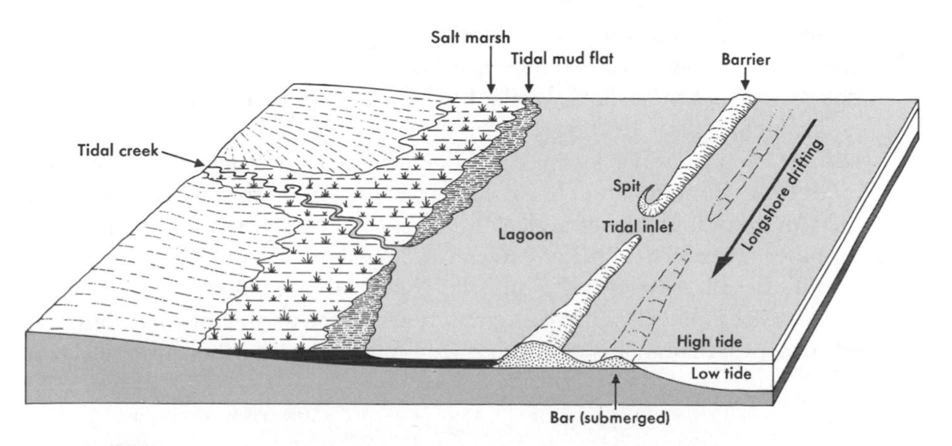

coastal plain that submerged, makes little difference in the sequential development of landforms on a flat coast. The principle element of a flat coast is a *barrier,* or beach ridge, separated from the mainland shore by a *lagoon* or marsh. Barriers protrude above normal high-tide level, which distinguishes them from submerged *bars.* Further descriptive classification of barriers is based on their form and position. Thus we have: *barrier spits,* barriers tied to the mainland at one end and building downdrift; *barrier islands,* separated like beads on a string by eroded *tidal inlets; baymouth barriers,* across the mouths of bays; *bayhead barriers,* enclosing a lagoon near the inland head of a bay; and others. Barriers are sometimes called offshore bars in the older literature, but this terminology should be avoided, for the term *bar,* like *reef,* traditionally has referred to a navigational hazard that is partially or entirely submerged.

Barriers are said to be found along fully one-third of the world's coasts and are therefore very important landforms. Formerly, two stoutly defended theories were proposed for their origin. One group of geomorphologists believed that barriers were built up by wave transportation of sand shoreward across the shallow ocean bottom. Surf action was supposed to stir up bottom sand, which was then piled onto the barrier by beach drifting and breaking waves. Another group, who believed that barriers are supplied by longshore drifting from an eroding headland somewhere upcurrent, denied that waves could move sediment shoreward over a sloping sea floor.

Curiously, both opposing viewpoints on the origin of barriers were strengthened by the intensive beach studies of World War II that were described in the Introduction to this book. New instruments and experimental techniques have established that sand can move onto beaches from a shallow offshore bottom under appropriate wave conditions. In fact, the volume of sand moved is much larger than was envisioned by the early defenders of this hypothesis of barrier formation. For example, during hurricane Audrey in June, 1957, the barrier along the western Louisiana coast was flooded over by 10 to 12 feet of water, and much of the barrier sand was washed either back into the marsh or offshore by the high waves of the storm. Within two years, less violent wave action had rebuilt the seaward profile of the barrier to the pre-storm form. No upcurrent sand supply was available, therefore all the sand had been moved shoreward from the shallow ocean floor, including enough to replace the sand that had been washed over the barrier and into the marsh behind it.

Other postwar studies have verified wartime observations that barriers are fed in part by longshore drifting. On the south shore of Long Island, New York, for instance, a chain of barrier spits and barrier islands extends westward about 115 miles from Montauk Point to Rockaway Point. Along this entire coast, sand derived from eroding cliffs of a glacial moraine at the eastern end of Long Island is moved westward by the prevailing longshore current. Samples taken at one-mile intervals westward along the Fire Island barrier showed progressive decrease of grain size, progressive increase in the degree of roundness and sorting, and progressive loss of heavier mineral grains. Furthermore, electron microscope pictures of the surfaces of individual sand grains showed that sand from the cliffs is distinctively fractured and sharp, whereas with longshore transport westward, the sand grains become smoother and better rounded, and lose the distinctive glacially produced markings in favor of characteristic abrasion features of water transport. Similar evidence of longshore drifting as a source of sand to barriers has been gained from many other coasts.

One of the problems of the traditional debate about the origin of barriers was that they were believed to protrude too far above sea level to be only the result of wave and current action. Therefore, it was widely concluded that barriers were former submarine bars that had been emerged by a slight relative lowering of sea level. Thus, the term "offshore bar," which barriers were formerly called, carried a connotation that the coasts fringed by these forms were in the process of emerging from the sea. This contention has been disproved by numerous studies of barrier and lagoon coasts. In fact, it now appears that barrier formation is favored by slow submergence, which keeps an open lagoon between the barrier and the mainland, and at the same time furnishes the waves with a supply of sediment from submerging, eroding headlands somewhere upcurrent. Barriers are built above sea level by high storm waves and wind. Most of the sand more than a few feet above sea level has been blown into dunes by onshore winds. On many large barriers, the dunes are fixed by grass and trees and have well-developed soil profiles.

How To Describe Coasts

The ultimate goal of coastal geomorphology is to describe a coast fully in terms of its past history. Explanatory description remains the most concise and complete technique for describing landscapes that has been devised, in spite of past abuses (Introduction, p. 2; Chapter 5, p. 82). If a person can summarize the nature of the coastal rocks that are undergoing change ("*structure*"), the *processes* of weathering and erosion that have caused and are causing the change, and the *time* during which the sets of structures and processes have interacted, he has fully described the coast. Explanatory description is far more satisfactory than a purely empirical description based on the present shape of the coastal profile (steep or flat) or the map view of the shoreline (rectangular, triangular, curved, straight, and so on).

The explanatory description of coasts, using structure, process, and time as key concepts, requires deducing a series of sequential landscapes. In this respect, describing coastal landscapes parallels the procedures used in the previous chapter for describing subaerial landscapes. However, because coasts are ribbon-like, two-dimensional strips at the edges of the land, their sequential evolution contrasts in two fundamental ways with the development of three-dimensional subaerial landscapes.

The first contrast is in the energy supply. Wave energy is fed into coasts in a horizontal plane, over the surface of the ocean. Sequential evolution is not controlled, then, by progressive loss of potential energy as base level is approached. Instead, the wave energy against the shore diminishes through time because a wider and wider wave-cut bench is formed. Energy decline and decreasing rate of erosion on a coast take place as a result of horizontal, rather than vertical, landscape evolution, and there is no definable limit to wave erosion short of the entire removal of the landmass.

The other feature of coastal landscapes that makes their evolution distinct from that of entirely subaerial landscapes is that a slight change in level of either land or sea will begin an entirely new sequence of coastal landforms. Changes of level may rejuvenate subaerial landscapes by introducing new potential energy into river systems (Chapter 5), but the sequential evolution of these landscapes is only interrupted, not terminated and begun anew with each slight uplift or lowering. Most coasts show evidence of repeated changes of level by land or sea, such as elevated wave-cut benches and wave-built terraces (together called *marine terraces*), estuaries (drowned river mouths), submarine terraces, or other similar features. The explanatory description of coasts usually involves decisions of how many separate episodes of landscape evolution are to be included in order to make the description complete.

Denied the unifying concept of gradual loss of potential energy by erosion toward an ultimate base level, we must seek other criteria on which to base an explanatory-descriptive scheme for coasts. *Time,* with the accompanying tendency toward equilibrium between processes and form, remains the same unidirectional factor in coastal evolution as it is in the sequential development of all landscapes. But, in place of degradation toward base level, we must substitute a pair of contrasting horizontal changes in coasts, *progradation* or *erosion,* and a pair of contrasting vertical changes, *emergence* or *submergence.* To describe a coast, then, we must be able to describe sequential changes through time that include episodes of progradation (outbuilding of the shore), erosion, emergence, and submergence. Each of the variables will be discussed in turn, then a general model will be presented.

Time

Time, or some measure of the progress toward equilibrium between coastal processes and coastal landforms, must be an essential part of any explanatory-descriptive scheme. Time in this sense is relative; a coast of easily eroded sediment might pass through a series of sequential forms to a mature, well-adjusted stage, while in the same time interval an adjacent coastal segment of resistant rock might barely show a trace of wave erosion. If we know the actual amount of time in years during which certain changes occur on a coast, we can calculate the rate of change per year, or the "tempo" of change. The tempo of erosion and deposition are related in part to the wave and current energy acting on the coast. The division of coasts into "high-energy" and "low-energy" environments is a useful procedure that implies either rapid or slow changes in the coastal landscape.

Coasts not only erode or prograde at various rates, but they also emerge or submerge at various rates. Most of the world's coasts are now submerging, but some Canadian and Scandinavian coasts that have been deglaciated recently are now emerging at a rate of several feet per century. Part of the northeastern Greenland coast has emerged 240 feet in the last 9,000 years; since during that same time, sea level has risen at least 100 feet, the total uplift of the land has been a very rapid 340 feet in 9,000 years. As a result of the great Alaskan earthquake in March, 1964, some parts of the Alaskan coast emerged as much as 33 feet in a few minutes, and along hundreds of miles of coast, instantaneous emergence or submergence of three to six feet was commonplace.

Progradation or Erosion

Progradation involves an increase of land area at the expense of the sea by deposition along a shoreline. Progradation requires an abundant supply of sediment from either river mouths or a shallow, sandy offshore ocean bottom, or rapid accumulation of organic debris. Progradation also implies a certain spectrum of energy on the coast, within which waves and longshore currents can distribute sediment laterally along the prograding segment of coast without abrading the sediment to such fine sizes that it is carried away from the coast in suspension. Prograding coasts may be *deltas* built at the mouths of rivers, *alluvial fan* coasts, *mangrove* and *coral-reef* coasts in warm tropical waters, *barrier* coasts, *marsh* coasts and others. Progradation can generally be recognized on maps or aerial photographs (Fig. 6–7) by multiple beach ridges, the older ones at the landward edge commonly fixed by vegetation. Deltas have a variety of forms, but all of them are conspicuous seaward bulges of the shoreline at the mouths of rivers.

Eroding coasts are most easily identified when they are backed by steep cliffs that show active mass-wasting. As long as waves and longshore currents erode the taluses at the foot of sea cliffs, the cliffs will continue to retreat, and the area of the sea gains on the area of land by erosion.

Eroding flat coasts are not as easy to recognize as steep, cliffy coasts, but any coast without a sandy beach at the shore must be suspected to be eroding. Another indication of erosional retreat of a coast is a *betrunked* river system, in which streams converge as though they were tributaries of a former master stream, but fail to join before they reach the coast. River valleys that end in waterfalls at the coast are good evidence that erosional retreat of the coast is proceeding more rapidly than subaerial stream erosion of the regional landscape.

On some coasts, multiple beach ridges are truncated by low wave-cut cliffs. On such coasts, former progradation has been replaced by erosion. This may

FIGURE 6–7 *A prograding coast in the Sea Islands of Georgia. Older beach ridges are covered with small trees as the shoreline is shifted seaward by new sand deposition.*

be part of a cyclic or seasonal change, or it may represent a permanent change in the sequential evolution of the coasts.

Submergence or Emergence

Submergence and emergence refer to relative changes of level of land and sea. Whether it is the absolute level of the sea that changes or the absolute level of the land is not significant. The terms must always be used only in the sense of relative changes of level. For instance, in recently deglaciated coastal areas, the land is rising because of the removal of a massive ice load. The sea may also be rising. If the land is rising faster, emergence results, but if either or both movements were to operate at a slightly different rate, submergence might result, even though both land and sea are actually rising.

Submerged coasts are often characterized by drowned river valleys, or *estuaries*. Chesapeake Bay, in the eastern United States, is an outstanding example of a large estuary. The Rio de la Plata, at the confluent mouths of the Rio Paraná and Rio Uruguay, between Buenos Aires, Argentina, and Montevideo, Uruguay, is another good example. So many of the world's great rivers enter the sea in estuaries that it is an inescapable conclusion that worldwide submergence has been a major event of recent geologic time.

Other signs of recent submergence are provided by well-defined river channels that cross shallow continental shelves. On the Sunda Shelf, a shallow ocean floor between Sumatra and Borneo in Indonesia, a well-defined dendritic valley system over 600 miles long grades northward to a former sea level about 300 feet below the present one. The drowned Hudson River channel can be traced across the continental shelf of the Atlantic coast of the United States to a drowned delta at a depth of about 250 feet. Such channels are easily identified on hydrographic charts, but of course are not revealed by topographic maps or aerial photographs of coastal regions.

Emerged coasts may reveal the relative movement of land and sea by emerged marine terraces, marine sediments covering the landscape above present sea level, wave-cut notches on cliffs above the reach of present waves, or emerged barrier ridges and sand bars on flat coasts. Emergence is generally easier to demonstrate than submergence, because the evidence for emergence is exposed to view, whereas much of the evidence for submergence must be gained by drilling or diving offshore. Older emerged marine features, however, may be eroded or weathered so that they are recognized only with difficulty.

A Graphic Scheme for Describing Coasts

All of the variables of coastal evolution: erosion, progradation, submergence, emergence, and time, can be represented on a single three-dimensional diagrammatic framework (Fig. 6–8). The origin of the figure can be any arbitrary

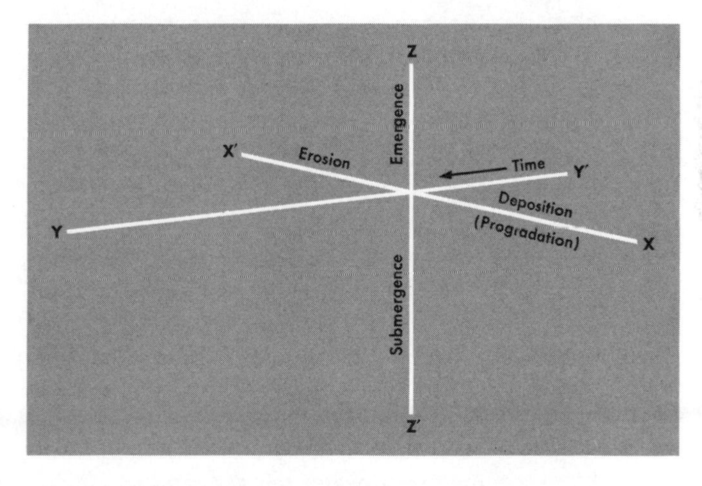

FIGURE 6–8 *A three-dimensional framework for the explanatory description of coasts.*

moment in time. The axes can be assigned any scales of time and length, either relative or absolute. Any point plotted within this framework represents the trends of change on a coast at one moment of time, and a line connecting a series of such points is a graphic representation of the coastal history. For example, a point in the upper foreground of Fig. 6–8 would represent a coastal segment that is both progaded and emerged. A straight-line trace of the point since "time zero" would show that this coast has been prograding and emerging for some time. The shoreline will have advanced seaward by the combined effects of emergence and progradation.

A significant feature of the graphic representation of coastal evolution is that it shows clearly how one variable of coastal change can be balanced by another. For instance, if a coast is emerging, but eroding at an appropriate rate, the shoreline will stay at the same position on a map. Similarily, on a submerging coast with active progradation, the shoreline might stay fixed on a map although the coastal region would change from a drowned upland to an ever-widening belt of abandoned beach ridges and dunes, or to a barrier island or reef separated from the mainland by an ever-widening lagoon. A three-dimensional representation of the coastal history reveals these changes.

An example of an explanatory description of an actual coast is provided by Fig. 6–9 and the following paragraphs. By collecting buried fresh-water peat

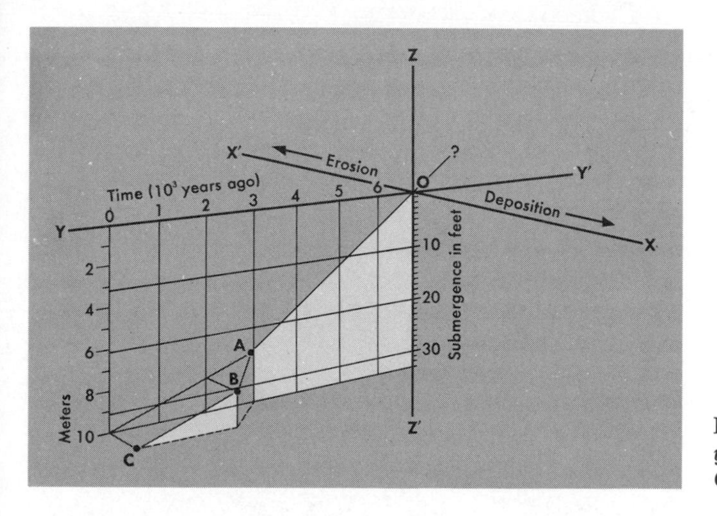

FIGURE 6–9 *The last 7,000 years of coastal evolution in Connecticut. (Bloom, 1965.)*

in bore holes to the bottom of tidal marshes along the Connecticut coast of the United States, and determining the age of the peat samples by the radiocarbon dating technique*, it has been possible to decipher the last 7,000 years of coastal evolution. When a coast submerges (Fig. 6–6), salt marshes migrate landward over fresh-water marshes and trees. Thus, by dating the buried peat and wood beneath the marshes, the rate of submergence can be established. From Fig. 6–9 we can see that from 7,000 years ago until about 3,000 years ago (segment 0 — A on the diagram) submergence was so rapid that very little erosion or deposition took place on the coast of Connecticut. The water, rising over the land at a rate of 0.6 feet per century, simply drowned the landscape before the weak waves on Long Island Sound could accomplish much erosion. Deposition was also not rapid enough to fill in the bays as fast as they were submerging.

About 3,000 years ago, the Connecticut coast abruptly decreased its rate of

* See Don L. Eicher, *Geologic Time,* p. 129.

submergence to about 0.3 feet per century, half of the earlier rate. Mud deposition in lagoons and estuaries exceeded this slower rate of submergence, and within an estimated 1,000 years, most of the former lagoons and estuaries were filled with sediment to the mid-tide level. The exposed tidal mud flats were soon occupied by salt-marsh grasses, which trapped more sediment and completely filled most former indentations of the coast with marshes. During this brief interval (Fig. 6–9, segment A–B), marsh deposition added about 43 square miles of new land along the 95-mile length of the coast. Subsequently, submergence and erosion (Fig. 6–9, segment B–C) have tended to cause the shoreline to retreat slightly, but new marsh growth and shoreline deposition have almost balanced the losses. The future of the Connecticut shoreline is not clear, for the tide-gage records of many Atlantic seaports in the United States show that submergence has been in progress for the last century at a rate of well over one foot per century. Man has built seawalls, dikes, and breakwaters along low parts of the coast, and natural sedimentation on the marshes has managed to keep pace with the accelerated submergence, but it is worth considering that houses or factories built on low coastal land will be subject to increasing hazards of storm flooding in the next decades, for there is no sign of decrease in the present rapid submergence rate.

This kind of explanatory description of a coast requires extensive study, and modern means of dating various events. Nevertheless, when it is completed the coastal landscape at any point in time can be described, and the sequential changes from one landscape to the next can be understood as the result of interactions between emergence, submergence, progradation, and erosion. The diagram can also be used to group coasts into broad categories such as those that have been submerged and eroded, emerged and prograded, emerged and eroded, and so forth. A brief examination of aerial photographs, maps, and charts is usually sufficient to classify coastal landforms in such broad groups.

7

Ice on the land

Typically, water on the surface of the Earth is in the liquid state. About 98 per cent of all the water is in the oceans. Most of the remaining 2 per cent is on the land, in the solid state as *glacier ice* (Fig. 1–3). Even this small fraction of the total water available on Earth is sufficient to cover entirely one continent (Antarctica), most of the largest island on Earth (Greenland), and many more square miles of high plateaus and mountains on other continents. Water in the solid state has thermal and physical properties very different from those in the liquid state. Therefore, on the 10 per cent of the land area of the Earth that is now ice-covered, the processes of weathering, mass-wasting, and erosion are unusual. The "climatic accident" of glaciation (Chapter 5) is unusual enough to require a special chapter in any geomorphology book.

We cannot simply ignore the glaciated regions of today as oddities not deserving of special treatment, for contemporary ice caps and valley glaciers are only the remnants of former continent-sized ice sheets that covered North America south to the Missouri and Ohio Rivers, and northern Europe as far south as the Netherlands, central Germany, Poland, and the western part of Soviet Russia (Fig. 7–1). Whereas today about 10 per cent of the land is ice-covered, during the last 2 million years or so there have been repeated episodes during which about 30 per cent of the land area was glaciated. About three-fifths of the formerly glaciated land is in North America, about one-fifth is in northern Europe, and the remaining one-fifth is widely scattered over the Earth in smaller areas. Every region in which glaciers are found today shows evidence that the glaciers were greatly expanded as re-

FIGURE 7–1 *Areas of present and former glaciers in the northern hemisphere. In the southern hemisphere, Antarctica, which is now almost completely ice-covered, had an even thicker ice sheet during the glacial maxima, valley glaciers in the Andes and New Zealand were formerly larger, and Tasmania supported a small ice cap where no glaciers exist today. (Compiled from Flint, 1957, and other sources.)*

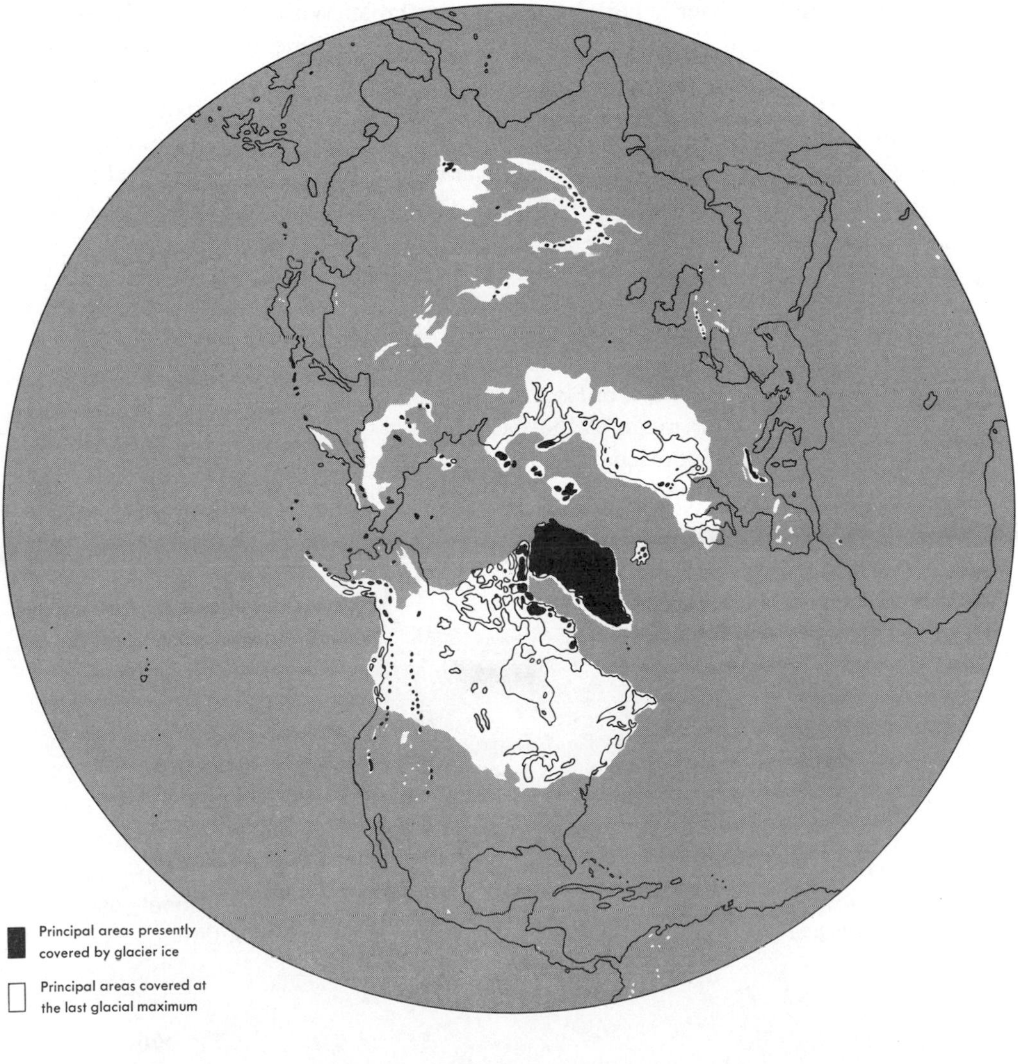

Principal areas presently covered by glacier ice

Principal areas covered at the last glacial maximum

cently as 10,000 to 20,000 years ago, and many regions that now have no glaciers show in their landscapes evidence of past glaciation.

The repeated glaciations during the last 2 to 3 million years that constitute the Pleistocene Epoch are only one aspect of climatic fluctuations over the entire Earth during this latest part of geologic time. The magnitude of the Pleistocene temperature changes is surprisingly small. The surface temperature of tropical oceans may have varied only about 6°C (11°F) between glacial and interglacial episodes. Although the temperature contrast between glacial and interglacial intervals may have intensified poleward, we need not visualize the Earth as gripped by intense cold during a glaciation.

For many species, life went on as usual from glacial to interglacial episode. Some groups of animals gradually migrated equatorward as the ice caps advanced, and dispersed poleward again as the climate warmed. Plants migrated too, by growing in favorable locations and gradually dispersing their seeds into new regions as former regions became inhospitable. With interesting but minor exceptions, the Pleistocene Epoch has not been a time of extinction of old forms of life and evolution of new forms. The name *Pleistocene* ("most recent life") was originally chosen by Sir Charles Lyell in 1839 to designate the latest interval of geologic time during which nearly all the modern species of plants and animals have been present. The Epoch is significant archeologically, for man is one of the modern mammals that characterize—in fact define—it.

Two separate themes intertwine throughout this chapter. The first concerns the geomorphic work of modern glaciers and their associated rivers of meltwater. The second concerns the work of the greatly expanded glaciers during the colder intervals of the Pleistocene Epoch. As well as we can judge, we are now in an interglacial episode of relatively small glaciers and high sea level. We do not know the cause of Pleistocene climatic fluctuations, but the periodicity and the similar range of the past temperature fluctuations suggest that future expansion of continental ice sheets is more probable than not. We still live in the Pleistocene Epoch, although for convenience, some chronologies recognize a "Recent" Epoch covering the time since the most recent melting of the North American and European ice sheets.

Snow, Ice, and Glaciers

Glaciology is the science of ice. It includes the study of ice crystals in high clouds, hail, and snow; frozen lake, river, and ocean water; and glacier ice. Glaciologists are also meteorologists, physicists, and geologists. Ice can be studied either as an easily deformed crystalline solid with a structure to be analyzed under a microscope, or as a geomorphic agent that erodes valleys and transports massive loads of rocky debris. What we know about glaciers has been learned from a wide field of experiment and observation.

Glacier ice primarily accumulates as snow on land where the mean annual air temperature is near freezing and where more snow falls in winter than can melt during the summer. The processes of converting snow to glacier ice include *sublimation, melting and refreezing,* and *plastic deformation.* Snowflakes are well known to be lacy, platy, skeletal ice crystals. Fresh snow is full of entrapped air, and may have a bulk specific gravity even lower than 0.1. That is, a volume of loose snow might weigh only one-tenth as much as an equal volume of water. Snowflakes readily sublime, or pass directly from the solid to the vapor phase, because of their very large surface area. Thus, old snowflakes lose their frilly margins and become more globular. If snowflakes are melted slightly and then refrozen, the high surface tension of water also tends to draw the original snowflakes together into equidimensional granules of ice. Meltwater often freezes onto snowflake nuclei, so that aging tends to increase the size of the ice granules.

Old snow, such as is found on sheltered mountain slopes in early summer, has the texture of very coarse sand. All traces of the original snowflake form is lost in this late-season *corn snow* or *buckshot snow,* familiar to avid skiers. The loose, granular mass is about half ice and half entrapped air, and has a bulk specific gravity of about 0.5.

Snow that has survived a summer melting season is called *firn* in German or *névé* in French. Both terms are widely used in English. Firn, or névé, is an intermediate step in the conversion of snow to glacier ice. It is granular and loose unless it has formed a crust. It represents the net positive balance between winter accumulation and summer losses.

As successive annual layers accumulate, the deep firn is compacted. The individual ice grains freeze together and the included air is either expelled or becomes enclosed as bubbles in the ice. By definition, when grains of ice are frozen together so that air is prevented from permeating through the mass, firn becomes glacier ice. The bulk specific gravity is usually about 0.8 by this stage of consolidation. The remaining air can be expelled only slowly by shearing and breaking or by recrystallization. Therefore, most glacier ice is a polycrystalline mass of frozen water plus a variable amount of air. Other components are dust and rock fragments that have fallen, washed, or been blown onto the ice surface, and rock that has been eroded from beneath the glacier. The bulk specific gravity of glacier ice varies from about 0.8 to about 0.9, which is approximately the specific gravity of pure, gas-free ice.

Ice is not a strong solid. Its normal crystal form is hexagonal, as we see when we examine snowflakes under a magnifying glass. Perpendicular to the vertical axis of an ice crystal is a *slip plane,* in which crystalline bonds are especially weak. Between any two layers of molecules in the crystal, movement parallel to the slip plane is easy. Ice crystals begin to deform measurably under a unidirectional pressure of slightly less than one atmosphere in excess of the confining or hydrostatic pressure. That is, if a specimen of polycrystalline ice is subjected to a unidirectional differential or shearing pressure of about 14 pounds per square inch, it will slowly but permanently deform by internal adjustments in the crystalline grains of ice. In a year, a load of about 14 pounds per square inch will flatten a small column of ice by about 30 per cent of its original length, with no melting involved.

Field experiments of glacier flow do not agree very well with laboratory studies of small ice specimens, probably because actual glaciers are not homogeneous and they include air bubbles and rock fragments. In theory, glacier ice buried by only a few tens of feet of overlying ice, firn, or snow should deform plastically, or *flow*. In fact, open crevasses, that demonstrate brittle cracking, are known to extend more than 100 feet downward into glaciers. Nevertheless, the laboratory experiments demonstrate that at great depth, glacier ice can deform as readily as if it were a very viscous liquid, and yet remain entirely solid and well below the melting temperature.

Pressure also lowers the freezing point of ice, but only very slightly. A confining pressure of 390 tons per square foot, equal to the maximum known thickness of the Antarctic ice sheet, lowers the freezing point only about 5.5°F (Fig. 2–3). *Pressure melting* in glacier ice is only important where very high pressures are concentrated, as for instance in the familiar demonstration of a weighted wire that cuts through a cold block of ice. The ice melts under pressure in front of the wire, and refreezes (or regelates) immediately behind it, so the wire cuts through the block of ice and yet the ice remains in one piece. The same principle is involved when you pack a snowball in your gloved hands. Within a mass of firn, pressure melting and regelation are common where two irregularly shaped grains are in contact at a point or on a thin edge. At such places, compacting forces are concentrated in very small areas, and melting results. The water immediately migrates to an adjacent area of lower pressure where it refreezes in crystalline continuity with an ice grain. By this process, as well as by other complex processes of recrystallization and migration of crystal boundaries, the crystalline texture of ice grows coarser as firn becomes glacier ice. Ice crystals an inch or more in diameter are common in glaciers.

Glacier Temperatures and Processes of Flow

Glaciers are landbound masses of moving, impure ice that form by a net accumulation of snow and frozen rain. They are often classified by their shape, as: (1) *valley glaciers,* confined between rock walls; (2) *piedmont glaciers,* lobate tongues spreading over the plains at the foot of glaciated mountain ranges; and (3) *ice caps* or *ice sheets,* masses of ice, convex to the sky, that bury the rocky landscape and flow radially outward under their own weight, independent of the underlying topography. This is a useful descriptive classification, but to understand the geomorphic work that glaciers do, a classification based on temperature is more useful.

Throughout the following discussion, you must keep in mind three physical properties of ice. First, the latent heat of fusion of ice is almost 80 calories per gram, which means that when a gram of ice melts, 80 calories are absorbed from the environment without a change of temperature; when a gram of ice freezes, 80 calories are released to the environment. Second, the temperature of coexisting ice and water is constant at any given pressure. Ice water at one atmosphere pressure remains at 0°C, or 32°F, until either all the water is frozen solid or the ice is entirely melted, regardless of the amount of the heat added to or removed from the mass during the coexistence of the two phases. Third, ice has a low thermal conductivity that is comparable to or lower than the conductivity of many rocks.

Glacier-Ice Temperatures

Glacier-ice temperatures closely approximate the mean annual air temperature at the places of accumulation, if that temperature is below freezing. Because ice is a poor conductor of heat, especially when it is full of entrapped air, each layer of ice tends to keep the temperature of the atmosphere at the time of snow accumulation. The temperature of each layer within the uppermost few feet of a glacier or snowfield depends on the time of year of each snowfall, but by the time ice has been buried about 30 feet, the seasonal temperature variations within the annual layers have averaged out, and the temperature of the ice or firn is the mean annual air temperature. Thus, when members of a United States research team inserted a thermometer into the wall of an ice pit only 30 feet deep at the South Pole, they acquired an important piece of climatic data. The mean annual temperature there is about −51°C (−60°F).

From the thermal properties of ice, we can visualize two basic types of glaciers. One type, the *polar* (or "cold") glacier, is below the melting temperature, and is solid ice. The other type, the *temperate* (or "warm") glacier, is at the melting temperature throughout, and interstitial water saturates the ice.

Polar glaciers are easiest to understand. The snow falls at temperatures far below freezing. With such cold air, the glacier stays thoroughly refrigerated at the surface. We must assume that geothermal heat (Chapter 1) flows into the base of a polar glacier at a rate comparable to the world annual average of about 40 calories per square centimeter, but this small amount of heat is simply conducted upward through the ice toward the cold surface, where it radiates into the still-colder stratosphere. Surely, the glacier becomes warmer with depth, just as other rocks of the Earth do, but in theory, the entire mass of a polar glacier is solidly crystalline, and is frozen firmly to the cold rocks beneath.

Temperate glaciers have more complex thermal conditions. Any glacier that experiences a summer melting season is likely to be warmed to the melting temperature to a considerable depth in spite of the low conductivity of ice. Suppose that a melt water pool on the glacier surface drains down into a crevasse in the ice. The water has always been in contact with ice, so its temperature cannot be above 0°C, yet each gram of meltwater carries 80 calories of latent heat into the glacier mass. If the ice at depth is colder, the water refreezes. One gram of water that freezes releases 80 calories to the surrounding cold ice, enough heat to raise the temperature of 160 grams of ice by 1°C (specific heat of ice = 0.5). A thick surface zone of water-saturated ice at the melting point is thus formed, not necessarily by conduction, but by refreezing of some of the penetrating meltwater.

A unique property of an ideal temperate glacier is that it becomes colder with depth, in spite of coexisting ice and water. Under the pressure of overlying ice, the melting temperature is slightly depressed. If ice at a temperature of 0°C is buried at great depth in a glacier, a small amount of the ice melts. This change of state by a small amount of ice draws heat from the remaining ice, and the temperature of the coexisting ice and water sinks below 0°C in proportion to the pressure. (The exact relationship between pressure and the melting temperature of ice is shown by Fig. 2–3.) This is the explanation of the pressure-melting phenomenon that was described earlier in the chapter. It means that in glaciers formed from an accumulation of ice with some meltwater near the surface, the pressure-melting temperature is likely to prevail throughout their thickness. Water is present in small quantities throughout the glacier, but the ice and water mixture is nevertheless slightly colder with increasing depth and pressure.

Temperate glaciers act as perfect insulators to geothermal heat. Because they are colder at the bottom than at the surface, heat will not flow upward through them. The average annual heat flow of 40 calories per square centimeter from the Earth under a temperate glacier can only be expended, then, by melting a layer of ice about 0.5 centimeter thick at the bottom of the ice. Thus, temperate glaciers ride on a film of water over their rock floors.

Most glaciers today, with the principal exception of the Antarctic ice cap, are probably temperate. The southern part of the Greenland ice cap receives heavy, wet snow, or even rain in the summer, from winds blowing inland from the warm Gulf Stream. The rain and wet snow probably keep it near the melting point except in the highest, coldest interior part of the ice cap. In addition to the Antarctic ice cap, however, the northern part of the Greenland ice cap is polar, as may be the glaciers of the northern Canadian islands. The great valley and piedmont glaciers along the Pacific Coast of Alaska and Canada are certainly temperate. One unfortunate gold miner in southern Alaska attempted to drive a mine through the lower part of a glacier to reach the rock face beyond. Digging went well until he broke into a pocket of water under considerable pressure, which quickly washed him, his wheelbarrow, and his pickaxe out into open air. He could have wished for a polar glacier.

Most glaciers have not had their temperature profiles determined. Considerable progress in glaciological studies was made during the International Geophysical Year, because it is widely believed that glaciers hold the key to many problems of climatic change, both past and future. One interesting sidelight of glacier studies concerns the technique of boring test holes in temperate glaciers. A deep hole can be melted simply by mounting an electrically heated copper tip on the end of a string of standard drill rods. A gasoline-powered generator on the surface provides the energy. A hole melted deep into a temperate glacier does not freeze closed again, for there is no cold ice to absorb the extra heat introduced by the "hot point." Strings of sensing devices can be lowered far into the interior of the glacier until plastic deformation squeezes the hole closed. No such easy technique will work for cold glaciers. Full-size rotary drills and large electrical generators are necessary, and because of the problems of accessibility, fuel supply, and the failures of mechanical equipment at below-freezing temperatures, very little drilling has been done on polar glaciers. The first drill hole through the Antarctic ice cap, at Byrd Station, was completed in February, 1968. Although the surface ice is at the mean annual air temperature of −28°C (−18°F) and the ice cap has been assumed to be polar, the drillers encountered water at 7,100 feet, at the bottom of the hole. Either the bottom ice is older than expected, and has absorbed more heat, or the local geothermal heat flow is abnormally high. Due to the basal water film, the ice cap is slipping over the rock floor at a rate of one inch per day.

How Glaciers Flow

The "flow" of glaciers is due to a combination of: (1) internal deformation in ice crystals, (2) melting and refreezing, (3) basal slip of the ice over rock, and (4) brittle fracture and faulting in the ice. A polar glacier probably flows only by internal plastic deformation of ice crystals and a minor amount of brittle fracture near the surface. It has been demonstrated experimentally that ice frozen to rock forms a bond that is stronger than the internal strength of individual ice crystals. Therefore, in theory at least, a cold glacier does not move by sliding over its bed, but by plastic deformation of the ice itself. One consequence of this form of motion is that polar glaciers erode and transport very little rock debris. The ice at the edge of the Antarctic ice sheet is almost free of rock fragments, suggesting that only slight erosion is in progress in the interior of the continent. Of course, when a rocky obstruction is broken away by the flow of cold ice, it is fragmented and transported until the ice melts or breaks off into the sea.

Temperate glaciers generally move faster than polar glaciers, because in addition to crystalline deformation and brittle fracture, the ice can deform by melting and refreezing, and by slipping over rock. Basal or bed slip is highly variable, but typically it accounts for about half of the total forward motion of many valley glaciers. Melting and refreezing under differential pressure is hard to separate from slip, because if a rock obstacle protrudes into the bottom of a temperate glacier, the excess pressure causes melting on the upstream side, and the water migrates downstream to the lee of the obstacle and refreezes. The glacier advances by "slipping" over the obstacle on a water film, but individual grains of ice are melted and refrozen to permit the slip. Basal slip in a temperate glacier seems to be determined by the bed roughness in a critical size range.

Temperate valley glaciers move downhill at rates generally between a few inches per day and five to six feet per day. On very steep slopes, or during the meltwater season when the bed seems to be lubricated by more water, they move several times as fast. Rarely, one part of a glacier will surge forward for a few weeks or months at a rate of several hundred feet per day.

Surges, or waves of rapid motion that move downstream through glaciers, are of particular interest because of the information they provide about the way glaciers flow. Typically, a glacier surge is triggered by an abrupt increase in accumulation on the upper part of the glacier, possibly caused by snow avalanches or exceptionally heavy seasonal snowfall. A wavelike bulge several feet high moves through the glacier at up to five times the velocity of the actual ice movement, much as an ocean wave moves through water. Because of the ease with which ice is deformed, it is assumed that the normal glacier motion before and after the passage of a surge is in response to a balanced set of forces, and the surge is the result of a single additional variable such as an abrupt increase in accumulated snow. In scientific research it is useful to be able to isolate single sets of causes and effects, and the surging responses of Alaskan glaciers to snow avalanches triggered by the 1964 earthquake are presently under intense study.

The surface layers of glacier ice do not deform in a plastic fashion, for the confining pressures are too low and the stresses too abrupt. Ice to a depth of about 100 feet is brittle and cracks into a maze of crevasses as it is carried along on the more plastic ice at depth. Near the lower terminus of a glacier, surface crevasses extend so deep into the ice that a basal slab of ice may become detached and cease to move, and active ice will ride forward over it on a *thrust fault* (Fig. 7–2).

Most glaciers have a high-altitude *zone of accumulation,* where accumulation exceeds losses, and a lower *zone of ablation,* where net loss of ice takes place (Fig. 7–2). Ablation includes melting, evaporation, loss by icebergs breaking off into water ("iceberg calving"), and other, minor losses such as by wind deflation. The lowest altitude of annual net accumulation on a glacier is marked by the *firn limit,* which is the equivalent of the *snowline* on the nearby mountain sides. In the zone of accumulation, new ice is constantly being added on top, so the general path of flow of an ice particle is down into the glacier, as well as downhill. In the zone of ablation, melting and other ablative losses lower the surface so rapidly that even though an ice particle may be following a downhill path, it eventually approaches the surface of the ice and melts.

Through the cross section of the glacier at the firn limit, all of the net annual

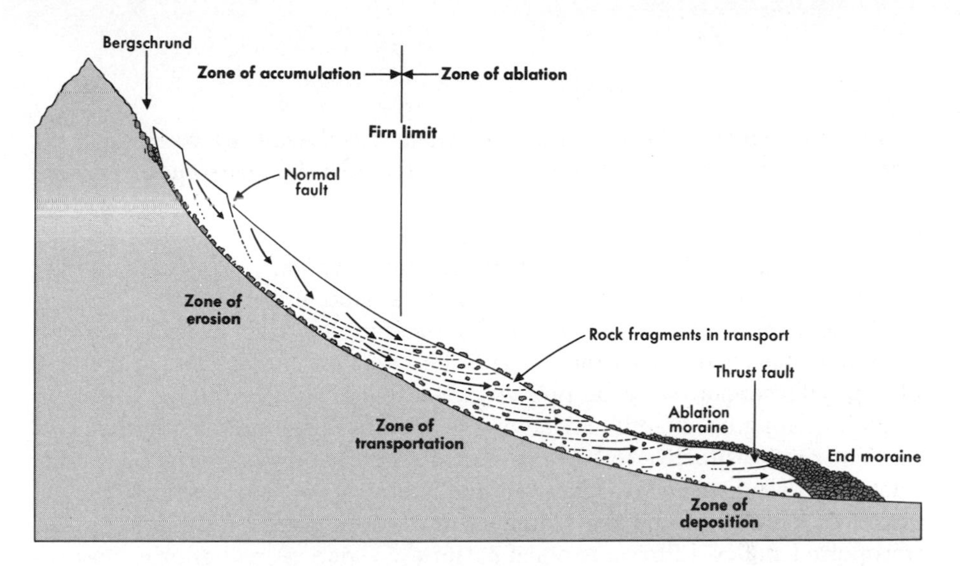

FIGURE 7–2 *Paths of motion in a valley glacier and regions of accumulation, ablation, erosion, transport, and deposition. The* bergschrund *is a characteristic crevasse at the head of a valley glacier, which traps much frost-wedged rock debris.*

accumulation from upstream must move to balance the net annual losses by ablation downstream. The flow beneath the firn limit is approximately parallel to the ice surface. Lacking equipment or time for a complete study of ice accumulation and ablation on a glacier, glaciologists can study the rate of flow at the firn limit in order to get a quick idea of the *regimen*, or dynamic status, of the glacier. Rapid flow at the firn limit implies both abundant snowfall upstream and rapid ablation below. This is the typical condition of a temperate valley glacier on a west-facing maritime slope in the mid-latitudes. Slow movement at the firn limit implies a sluggish regimen, with both slow accumulation and minor losses by ablation. The Antarctic ice cap is probably the most sluggish in the world. The firn limit there is virtually at sea level at the edge of the continent, and iceberg calving is the only major form of ablation. Over the entire continent, annual precipitation averages less than the equivalent of five inches of water. After decades of study, it is still not known whether the Antarctic ice cap is enlarging or shrinking, for the rates of both accumulation and ablation are too low to be measured with confidence.

Glacier Erosion and Transport

A geologist is primarily interested in ice as an agent of erosion, transportation, and deposition of rock, or as a shaper of landscapes. We need to apply what we have learned about ice deformation and glacier motion to geologic problems. At this point, the subject matter changes from glaciology, the study of ice, to *glacial geology*, the study of the geologic work of glaciers.

Ice erodes rock in many of the same ways that flowing water erodes rock. Both can exert great force against an obstruction, and break pieces from it as they flow past. Both carry rock fragments as tools that can abrade the rock surface over which they move. Water has the advantages of greater velocity of flow and greater turbulence, but ice has the advantages of greater rigidity and the ability to melt and refreeze as it flows around obstacles.

Because of its slow rate of plastic deformation, ice tends to hold rock particles in contact with each other or with the rock floor. A rock tool at the base of a glacier cannot easily be pressed upward into the ice, and is forced to scratch a long groove on the underlying rock and to be ground flat in the process. The most distinctive features of glacial erosion and transportation are the scratches (called *striations* if they are fine hairlines) and grooves made on glaciated rock surfaces, and the flattened, or *faceted,* polished surfaces on ice-transported stones. Either a regional pattern of striations and grooves on bedrock surfaces, or faceted, striated pebbles and boulders of exotic rock types are sufficient evidence to infer former glaciation in an area not now ice-covered. It is true that numerous other natural and artificial agents can abrade bedrock, such as avalanches, lava flows, buffalo hooves, steel sled-runners, mudflows, floating icebergs scraping the sea floor, and even streams carrying bedloads of sharp sand, but none of these produce striations and grooves of such persistence in form and direction. A regional pattern of striations and grooves on bedrock surfaces probably constitutes the best single proof of former glaciation.

Stones transported in glacier ice develop flattened or gently curved facets that intersect at blunt corners and edges. Whereas water transport tends to develop better-rounded stones, glacial transport tends to develop angular shapes with flattened faces. Commonly, stones transported in glaciers have surfaces curved like the sole of an old shoe, as though the stone had been rocked to and fro against a grinding wheel. Some stones have a distinctive "flatiron" shape, with a blunt point at one end formed by the intersection of two or more facets.

Abrasion during glacial transport provides an abundance of fine-grained, mechanically crushed rock fragments. Most of the larger fragments carried by glaciers seem to be derived by the process of *plucking,* or *quarrying.* This is a process unique to temperate ice. As meltwater migrates toward the downstream, or lee, side of the obstruction and refreezes there, any loose fragments of rock on the lee slope are frozen into the new ice, and are carried away from the exposure. Heavily jointed crystalline rocks and thin-bedded sedimentary rocks are particularly affected by plucking. It has been demonstrated that glacial plucking is quantitatively more important than glacial abrasion on suitable rock types (Fig. 7–3).

Valley glaciers receive a large portion of their rock debris from the valley walls. Frost wedging is very active in areas with temperature and precipitation patterns conducive to glacier formation, and much of the debris that falls onto

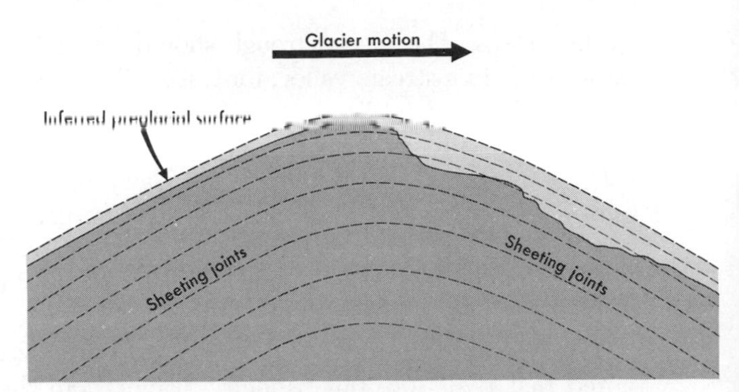

FIGURE 7–3 *Relative effectiveness of glacial abrasion and plucking in New England. If the sheeting joints follow the original shape of the hill, much more rock has been removed from the downstream end of the hill by glacial plucking than from the upstream end by abrasion. (After Jahns, 1943.)*

a valley glacier is unweathered except for having been broken free by repeated freezing and thawing. Debris that falls onto the zone of accumulation of a valley glacier may be carried deep into the ice, perhaps even to the rock floor beneath, before it moves back up to the surface at the lower end of the glacier (Fig. 7–2). Stones that travel such a path will likely have the distinctive markings of glacial erosion. Much of the debris that falls on a valley glacier stays near the surface, however, and is carried frozen in the brittle surface layer of ice. This material rarely shows distinctive marks of glacier transport. In fact, it is often indistinguishable from slide rock or landslide debris.

Erosion by temperate valley glaciers results in a distinctive landscape that clearly rates high among the most impressive landscapes on Earth. The Alps of Europe are particularly well known for their striking glaciated landscapes, and give their name to the assemblage of landforms that compose *Alpine scenery.* Since most Alpine glacial landforms are erosional, this is an appropriate point to describe glacially eroded landscapes in general.

The basic unit of Alpine landscape is the *U-shaped trough* (Fig. 7–4). This distinctive valley form has straight, steep side walls, blunted or *truncated spurs,* and a step-like long profile often consisting of a series of *rock basins* separated

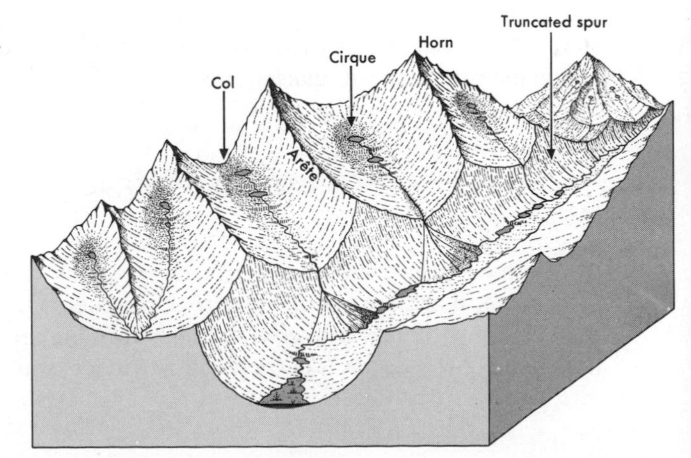

FIGURE 7–4 *An Alpine landscape. The result of glacial erosion.*

by low steps. The glacial trough should be compared to a stream channel, rather than to a stream valley, for it is really the channel of a river of ice. Glacial troughs head in *cirques,* semicircular basins that form in the zone of accumulation. Adjacent cirques and glaciated troughs commonly intersect in *cols,* or passes, through knife-edged, sawtoothed ridges called *arêtes.* High pyramidal peaks along the arêtes are called *horns.* The Matterhorn in the Swiss Alps is regarded as the type example of a glaciated horn, but dozens of lesser peaks of similar form stand at the crests of glaciated mountains. Glacial troughs drowned by the sea are called *fjords.*

Glaciers erode troughs proportional to their size, just as Playfair long ago noted that rivers do. But tributary glaciers join the main valley stream with their ice surfaces at a common level, therefore the rock floors of tributary glacial troughs commonly are far above the floor of the master trough. These *hanging valleys* are another of the distinctive landforms of Alpine scenery. Streams from the hanging valleys cascade or fall into the valley below. Sutherland Falls, in New Zealand, drops 1,904 feet almost vertically from a rock-basin lake in a hanging valley to the floor of the main valley. It is one of the highest waterfalls in the world.

A smaller-scale erosional feature common to both the Alpine landscape and areas glaciated by ice sheets is the *glaciated rock knob.* Any glaciated bedrock hill is likely to be polished and abraded on the upstream side and plucked or quarried on the downstream side. Swarms of such asymmetric hills cover glaciated uplands and valley floors alike. By their strong asymmetry, glaciated rock knobs tell the regional direction of ice movement even more clearly than glacial striations and grooves. In France, glaciated rock knobs are called *roches moutonnées,* apparently because the abraded upstream faces have the smoothly rounded appearance of the woolen wigs worn in the seventeenth and eighteenth centuries. Another name given to a landscape of glaciated rock knobs is *stoss-and-lee topography,* emphasizing the contrasting appearance of the smooth, abraded, gentle upstream slopes (stoss slopes, from the German verb *stossen,* to push or thrust) and the hackly, broken, steep, lee slopes. Glaciated rock knobs are apparently stable landforms under ice. Their shape seems to represent a kind of streamlining that minimizes glacial erosion. If we understood the dynamics of their origin, we would necessarily know much more about the mechanics of glacier erosion.

Glacier Deposition

We have seen that glaciers abrade or pluck the rock surface over which they move, that they pry or break rock fragments from buried ledges, and that they transport debris delivered to them from higher rock slopes by mass-wasting. Some of the debris is carried within the ice; some is carried on the surface.

Some rock waste is ground to fine powder; some is carried along scarcely modified by glacial erosion; some is sorted and rounded by flowing water. The end result is an extremely heterogeneous rock material called *glacial drift.*

It is important to distinguish clearly the material carried by glaciers, or the drift, from the landforms which are built of the drift. Three terms are restricted to designate certain kinds of glacial drift; these terms apply to materials, not landforms. They are: (1) *glacial till,* the nonsorted, nonstratified debris typically deposited directly from glaciers; (2) *ice-contact stratified drift,* material partly water-sorted and crudely stratified, deposited adjacent to melting ice; and (3) *outwash,* fluvial sediment deposited away from melting ice by meltwater streams. Each of these materials is associated with certain depositional landforms, but since many landforms contain more than one kind of drift, terms that apply to landforms should not be confused with terms that designate the various kinds of drift.

The descriptive terminology applied to glacier-depositional landforms is complicated by the practice of using the same term for a feature on, in, or beside an active modern glacier and a feature formed in a similar way from a glacier that has now melted and vanished. *End moraines,* for example, are ridgelike mounds of drift at the downstream terminuses of modern glaciers, but they are also ridgelike mounds of drift across Ohio, or Germany, or anywhere else, that mark former ice margins.

Most glacier-depositional landforms accumulate during the melting of glaciers and the retreat of the ice front. As glaciers thin by ablation, especially frontal melting, the rock debris carried in them is either deposited directly from the ice or carried some distance by meltwater and deposited. The history we deduce from glacial deposits is only the history of the final stages of ice melting.

Either shape, or material, or both, determines the name given to glacier-depositional landforms. *Moraine* is a very general term for a landscape constructed of drift (Fig. 7–5). If moraines show strongly linear or arcuate ridges that outline former ice margins, they are called *end moraines.* Lacking distinctive linear elements, the moraine landscape is simply called *ground moraine.*

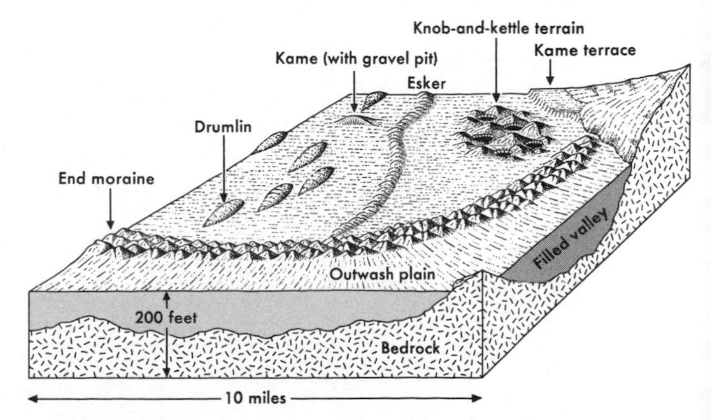

FIGURE 7–5 *Drift landscape. The entire surface is constructed from glacial drift except for small areas in which bedrock hills protrude.*

A moraine may have only slight relief, and slopes of no more than a few per cent, or it may be an irregular, hummocky maze of drift mounds separated by closed basins where ice blocks were buried in the drift, and then melted. These ice-block depressions, or *kettles,* are good evidence that some ice was still in the area at the time a moraine landscape was built, and a chronologic sequence of deglaciation can sometimes be interpreted from evidence such as kettles.

A moraine is generally formed of either till or ice-contact stratified drift. Some end moraines are composed entirely of till, and seem to represent either a slump of debris that piled against the ice margin, or a forward thrust of ice that "bulldozed" the ridge of drift. Other end moraines are composed almost entirely of ice-contact stratified drift, from which it can be inferred that the ice front stayed in one position for a time while forward motion just balanced ablation. Meltwater, flowing from the surface of the ice, washed drift into crevasses and banked it against the ice front. Then, after the ice melted, ridges of ice-contact stratified drift remained to mark the former ice-front position.

There are several especially distinctive landforms built of ice-contact stratified drift. One is the *kame,* a roughly conical or flat-topped hill, usually composed of water-laid sand and gravel, but with included masses of till. Kames apparently form in holes or at the intersection of large crevasses in melting ice. Drift is washed or slumps into such depressions and the water drains away through the ice. When the last ice has melted, the former depression becomes a mound, reflecting by its shape the disposition of ice masses around it. Kames are widely sought as sources of construction gravel, and in all populated regions they will be partially excavated and their interior character exposed. Gravel pits in kames reveal endless examples of the chaotic sedimentary structures that characterize ice-contact stratified drift. A variant of the kame is the *kame terrace,* a depositional bench along a valley wall; this forms when sediment fills the depression along the margin of a melting ice mass that still occupies the center of a valley.

Even more distinctive than the kame is the *esker,* a sinuous ridge of water-laid drift that may be as much as a hundred feet high and may snake across a moraine landscape for dozens of miles. Eskers probably form in former ice-walled river channels within or beneath melting ice. The meltwater streams apparently carry such heavy sediment loads that they aggrade their channel to maintain a proper slope. Eskers record the sinuosity of the former ice-walled channels, only slightly modified by mass-wasting of the esker sides that followed the removal of the retaining ice walls. Eskers, like kames, are eagerly sought as sources of well-washed sand and gravel. Many eskers near populated areas have been entirely consumed by gravel operations.

Where sediment-laden meltwater flows beyond the last blocks of melting ice, it becomes an outwash stream. All the "laws" of hydraulic geometry apply to outwash streams, although flow characteristics are somewhat special. Meltwater discharge is usually most rapid in late afternoon, when the daily heat has produced much surface runoff. By dawn, an outwash stream may be reduced to a slight trickle of ice water, only to swell to a torrent again by midday. Under these conditions of discharge, and carrying a heavy sediment load, outwash streams usually have braided channels, and commonly build up their valley floors with many feet of alluvium in order to maintain their flow. Outwash streams thus build *outwash plains* across open land, or *valley trains* on the floors of valleys. The slopes of outwash plains and valley trains may be quite steep, up to 300 feet per mile in some places. Such slopes are more typical of alluvial fans or pediments than of stream channels. There are close analogies between the flash floods of deserts and the daily peak discharges of meltwater streams.

One distinctive group of moraine landforms defies classification as either erosional or depositional in origin. Though formed of glacial drift, usually till, they have smooth streamline forms that argue for shaping by erosion. They are collected under the non-genetic title of *streamline molded forms* to imply origin by both erosion and deposition under moving, probably temperate, ice. The best-known streamline molded forms are the *drumlins.* These are hills of various sizes, but all distinctly elliptical in plan view and very much like the inverted bowl of a teaspoon in profile. Commonly their crest is near the upstream end and they have a teardrop-like tail, but some are not so notably asymmetric. They occur as *drumlin swarms* in southern Wisconsin, in New York State south of Lake Ontario, in eastern Massachusetts and New Hampshire, in Nova Scotia, in Ireland, and in many other glaciated regions.

Drumlins must be molded under moving ice, for their streamline form is obviously a response to dynamic conditions. Some have rock cores and seem to have formed by till from the base of the glacier being plastered over an obstacle. Others have no evidence of any primary obstruction that acted as a core. It has been noted that they are systematically spaced either in ranks and files or *en echelon,* and some observers have suggested that their form is related to some kind of standing wave or harmonic irregularity in the ice flow over an area. Others have suggested that divergent flow at the front of an ice lobe might promote deposition beneath the glacier margin. The fact is that no one has ever seen drumlins or other streamline molded forms in the process of formation, nor is anyone ever likely to see them. But these forms tell us the direction of ice flow as clearly as do stoss-and-lee landscapes, and they record a stage of flowing, active ice rather than the more typical stage of stagnation and melting recorded by most depositional landforms.

Other streamline molded forms include the *fluted moraine,* with a systematic topographic corrugation parallel to ice flow, and the *crag-and-tail,* a resistant rock knob with till deposited in the downstream shadow region. Both fluted moraine and crag-and-tail can be thought of as variants of the ideal drumlin form.

Pleistocene Climatic Change

Glaciation is a dramatic "accident" that befalls a landscape. The changes in valley and hill shape that result from overriding glacier ice, and the drift that is left behind when the ice melts, are so obvious that the limit of glaciation can usually be determined to within a mile or so, and frequently even more precisely. Since about 1840, when Louis Agassiz published a little book in which he publicized the concept of a great "ice age," glacial geologists have mapped the glacial drift of the continents, and sorted it into a stratigraphic succession that records at least four advances and retreats of ice caps into mid-latitudes during the Pleistocene Epoch.

The latest glacial advance apparently began about 70,000 years ago, and after several pulses, culminated about 20,000 years ago in North America and Europe. Then, with a margin in general retreat from excessive ablation, but with intervals of temporary re-expansion, the continental ice caps shrunk to borders broadly defined by the Canadian-U.S. border in eastern North America, and by the Baltic Sea around Sweden. At this stage, about 10,000 years ago, the climate seems to have grown warm enough again so that the remaining deglaciation was very rapid, perhaps even catastrophic in some regions. By 6,000 years ago nearly all the ice of the two great former ice sheets had melted, and for the next few thousand years, the Earth seems to have been warmer and drier than at present. About 1,000 B.C., a slow, fluctuating tendency toward cooler climates began. Since the mid-nineteenth century the trend has been reversed, and historical records prove that the climate has been warming up, at least in middle North America and Europe, but this is at least in part due to the exceptional combustion of coal and oil that has marked the industrial revolution. Lacking a general theory for climatic change, we cannot predict the future; but basing a prediction on the history of Pleistocene glaciation, we can say that renewed growth of continental ice sheets within the next 100,000 years seems highly probable.

Glaciation is only one aspect of the climatic fluctuations of the Pleistocene Epoch. Ice now covers 10 per cent of the land; at each glacial maximum it covered about 30 per cent. What about the 70 per cent that was not glaciated?

As continental ice caps expand, sea level must fall, for the hydrologic cycle is a closed system (Chapter 1). To judge by the best estimates of the present volume of the Antarctic and Greenland ice caps, if all the ice were to melt, sea level would rise about 200 feet, flooding a large portion of the most densely inhabited lands of the Earth. By comparison, estimates of the volume of the former North American and European ice sheets suggest that during the latest full glacial interval, about 20,000 years ago, sea level was 350 to 400 feet lower than at present. The consequences of the postglacial rise of sea level on the evolution of coasts was discussed in the previous chapter. Most of the shallow continental shelves of the world are covered with relict fluvial sediment that is at most thinly veneered by postglacial marine deposits. Over much of the submerged shelf areas, ancient sediments are still exposed, and only near the shore has marine deposition buried the glacial-age relict deposits on many shelves. Postglacial shoreline erosion has barely begun.

The postglacial rise of sea level has drowned or alluviated the mouth of every river that enters the sea. We cannot see an example of an old-age erosional land surface near sea level, for all were dissected during glacially lowered sea level, and their valleys are now being filled with alluvium.

When the North American and European ice sheets were fully expanded, the other climatic zones seem to have been compressed toward the equator. The Mediterranean and North African regions were cooler and wetter, as is indicated by well-developed drainage networks that now lie dry in the deserts, and by tree-pollen grains in sediments that indicate former woodlands where trees do not grow today.

The dry southwestern region of the United States was cooler and wetter during glacial intervals, too. During the periods of greatest glacial extent permanent lakes filled many present playas. Trees grew several thousand feet lower on mountain slopes, and grassland soils formed where Aridisols now form. The effects of these climatic and vegetational changes on the evolution of the pediment landscape is not yet understood. Apparently, some pediments were dissected by permanent stream flow during more humid intervals of the Pleistocene Epoch. Arid regions may have experienced more rain, but not enough to promote ground cover by vegetation. Under these conditions, erosion and deposition may have been accelerated.

Dry landscapes on the Earth evolve so slowly that many of their landforms are relict from former wetter or drier, cooler or hotter climatic conditions. Parts of the dry Australian landscape, for instance, seem to record three or four cycles of increased and decreased precipitation, during which, alternately, soils formed and wind-blown sediments accumulated. Perhaps when a technique has been found to date the Australian climatic changes, they will prove contemporary with comparable changes in the Northern Hemisphere.

It should be obvious from this brief review of Pleistocene climatic change that only the very newest landscapes can be regarded as the product of the present environment. Any landscape more than a few thousand years old must have formed under more than one set of climatic and vegetational conditions. Recall the variety of weathering, soil-forming, mass-wasting, and erosional processes that characterize each climatic region, and try to envision the result of superimposing a new set of processes on a landscape that has been evolving under another set. The enormous variety and complexity of modern landscapes, which make our sight-seeing trips so rewarding, also make geomorphology one of the most intriguing branches of the Earth Sciences.

Suggestions for further reading

Chapter 1
Energetics of the Earth's surface

Gates, D. M., 1963, The energy environment in which we live: American Scientist, v. 51, p. 327–348.

Nace, R. L., 1960, Water management, agriculture, and ground-water supplies: U.S. Geol. Survey Circ. 415, 12 p.

Chapter 2
Rock weathering

Goldich, S. S., 1938, A study in rock-weathering: Jour. Geology, v. 46, p. 17–58.

Keller, W. D., 1957, Principles of chemical weathering: Columbia, Mo., Lucas Bros. Publishing Co., 111 p.

Simonson, R. W., 1962, Soil classification in the United States: Science, v. 137, p. 1027–1034.

Chapter 3
Rock fragments in motion

Eckel, E. B., ed., 1958, Landslides and engineering practice: Nat. Acad. Sci.—Nat. Res. Council Pub. 544, 232 p.

Kerr, P. F., 1963, Quick clay: Sci. Amer., v. 209, no. 5, p. 132–142.

McDowell, Bart, and Fletcher, J. E., 1962, Avalanche! 3,500 Peruvians perish in seven minutes: National Geographic, v. 121, p. 855–880.

Chapter 4
Streams and channels

Leopold, L. B., and Langbein, W. B., 1966, River meanders: Sci. Amer., v. 214, no. 6, p. 60–70.

Leopold, L. B., Wolman, M. G., and Miller, J. P., 1964, Fluvial processes in geomorphology: San Francisco, W. H. Freeman, 522 p.

Mackin, J. H., 1948, Concept of the graded river: Geol. Soc. America Bull., v. 59, p. 463–512.

Chapter 5
Life history of landscapes

Cotton, C. A., 1947, Climatic accidents in landscape-making: New York, Wiley, 354 p.

Cotton, C. A., 1948, Landscape as developed by the processes of normal erosion (2nd ed.): Christchurch, N. Z., Whitcombe and Tombs, 509 p.

Schumm, S. A., 1963, The disparity between present rates of denudation and orogeny: U.S. Geol. Survey Prof. Paper 454-H, 13 p.

Chapter 6
The edges of the land

King, C. A. M., 1959, Beaches and coasts: New York, St. Martin's Press, 403 p.

Shepard, F. P., 1963, Submarine geology (2nd ed.): New York, Harper and Row, 557 p.

Zenkovich, V. P., 1967, Processes of coastal development: English edition ed. by J. A. Steers and C. A. M. King, trans. by D. G. Fry, New York, Interscience-Wiley, 738 p.

Chapter 7
Ice on the land

Flint, R. F., 1957, Glacial and Pleistocene geology: New York, Wiley, 553 p.

Kamb, Barclay, 1964, Glacier geophysics: Science, v. 146, p. 353–365.

Sharp, R. P., 1960, Glaciers: Eugene, Oregon, Oregon State System of Higher Education, Condon Lecture, 78 p.

Literature cited

Berkey, C. P., and Morris, F. K., 1927, Geology of Mongolia: New York, Am. Museum Nat. Hist., 475 p.

Bloom, A. L., 1965, The explanatory description of coasts: Zeitschr. Geomorph., v. 9, p. 422–436.

Bridgman, P. W., 1911, Water, in the liquid and five solid forms, under pressure: Am. Acad. Arts Sci. Proc. (Daedalus), v. 47, p. 439–558.

Bryan, Kirk, 1922, Erosion and sedimentation in the Papago country, Arizona: U.S. Geol. Survey Bull. 730-B, p. 19–90.

Bunting, B. T., 1961, The role of seepage moisture in soil formation, slope development, and stream initiation: Am. Jour. Sci., v. 259, p. 503–518.

Cotton, C. A., 1958, Geomorphology, 7th ed.: Christchurch, N. Z., Whitcombe and Tombs, 505 p.

Davis, W. M., 1902, Base-level, grade, and peneplain, *in* Geographical essays (Reprint): New York, Dover Publications, p. 381–410. [1954]

Emery, K. O., 1968, Relict sediments on continental shelves of world: Am. Assoc. Petroleum Geologists Bull., v. 52, p. 445–464.

Feth, J. H., Roberson, C. E., and Polzer, W. L., 1964, Sources of mineral constituents in water from granitic rocks, Sierra Nevada, California and Nevada: U.S. Geol. Survey Water-Supply Paper 1535-I, 70 p.

Gilbert, G. K., 1880, Contributions to the history of Lake Bonneville: U.S. Geol. Survey, Second Ann. Rept., p. 167–200.

Gilbert, G. K., 1909, The convexity of hilltops: Jour. Geology, v. 17, p. 344–350.

Griggs, D. T., 1936, The factor of fatigue in rock exfoliation: Jour. Geology, v. 44, p. 783–796.

Hansen, W. R., 1965, Effects of the earthquake of March 27, 1964, at Anchorage, Alaska: U.S. Geol. Survey Prof. Paper 542-A, 68 p.

Hopson, C. A., 1958, Exfoliation and weathering at Stone Mountain, Georgia, and their bearing on disfigurement of the Confederate Memorial: Georgia Mineral Newsletter, v. 11, p. 65–79.

Houghton, H. G., 1954, On the annual heat balance of the northern hemisphere: Jour. Meteorology, v. 11, p. 1–9.

Jahns, R. H., 1943, Sheet structure in granites: its origin and use as a measure of glacial erosion in New England: Jour. Geology, v. 51, p. 71–98.

Judson, Sheldon, and Ritter, D. F., 1964, Rates of regional denudation in the United States: Jour. Geophys. Res., v. 69, p. 3395–3401.

Kessler, D. W., Insley, H., and Sligh, W. H., 1940, Physical, mineralogical, and durability studies on the building and monumental granites of the United States: U.S. Nat. Bureau of Standards Jour. Research, v. 25, p. 161–206.

Kiersch, G. A., 1964, Vaiont reservoir disaster: Civil Engineering, v. 34, p. 32–39.

Langbein, W. B., and Leopold, L. B., 1964, Quasi-equilibrium states in channel morphology: Am. Jour. Sci., v. 262, p. 782–794.

Leopold, L. B., and Maddock, Thomas, 1953, Hydraulic geometry of stream channels and some physiographic implications: U.S. Geol. Survey Prof. Paper 252, 57 p.

McGee, W J, 1897, Sheetflood erosion: Geol. Soc. America Bull., v. 8, p. 87–112.

Munk, W. H., and Sargent, M. C., 1954, Adjustment of Bikini Atoll to ocean waves: U.S. Geol. Survey Prof. Paper 260-C, p. 275–280.

Playfair, John, 1802, Illustrations of the Huttonian theory of the Earth (Facsimile Reprint): New York, Dover Publications, 528 p. [1964]

Schumm, S. A., 1960, The shape of alluvial channels in relation to sediment type: U.S. Geol. Survey Prof. Paper 352–B, p. 17–30.

Sharpe, C. F. S., 1938, Landslides and related phenomena (Reprint): Paterson, N.J., Pageant Books Inc., 137 p. [1960]

Shelton, J. S., 1966, Geology illustrated: San Francisco, W. H. Freeman and Co., 434 p.

Soil Survey Staff, 1960, Soil classification, a comprehensive system—7th approximation: U.S. Dept. Agric. Soil Conservation Service, Washington, D.C., 265 p.

Troeh, F. R., 1965, Landform equations fitted to contour maps: Am. Jour. Sci., v. 263, p. 616–627.

Index